Molecular Virology

Second Edition

THE MEDICAL PERSPECTIVES SERIES

Advisors:

Oncogenes and Tumor Suppressor Genes
Cytokines
The Human Genome
Autoimmunity
Genetic Engineering
Asthma
DNA Fingerprinting, 2nd Edn
Molecular Virology, 2nd Edn
HIV and AIDS
Human Vaccines and Vaccination
Antibody Therapy
Antimicrobial Drug Action
Molecular Biology of Cancer
Antiviral Therapy
Understanding Gene Therapy

Forthcoming titles:

Molecular Diagnosis

Molecular Virology

Second Edition

David R. Harper
Department of Virology, Medical College of St Bartholomew's Hospital, London, UK

with a contribution from:
Paul R. Kinchington
University of Pittsburgh, Pittsburgh, PA, USA

First published 1994 (1 872748 57 0)
Second edition 1998 (1 85996 246 7)
Reprinted 1999

A CIP catalogue record for this book is available from the British Library.

ISBN 1 85996 246 7

BIOS Scientific Publishers Ltd
9 Newtec Place, Magdalen Road, Oxford OX4 1RE, UK
Tel. +44 (0)1865 726286. Fax +44 (0)1865 246823
World Wide Web home page: http://www.bios.co.uk/

Published in the United States of America, its dependent territories and Canada by Springer-Verlag New York Inc., 175 Fifth Avenue, New York, NY 10010-7858, in association with BIOS Scientific Publishers Ltd.

Published in Hong Kong, Taiwan, Singapore, Thailand, Cambodia, Korea, The Philippines, Indonesia, The People's Republic of China, Brunei, Laos, Malaysia, Macau and Vietnam by Springer-Verlag Singapore Pte. Ltd, 1 Tannery Road, Singapore 347719, in association with BIOS Scientific Publishers Ltd.

Production Editor: Priscilla Goldby
Typeset by Marksbury Multimedia Ltd, Midsomer Norton, Bath, UK.
Printed by BPC Information Ltd, Exeter, UK.

Contents

Abbreviations **vii**

Preface **xi**

1 Virus structure and replication **1**

What is a virus? 1
Virus classification 2
Virus structure 8
Virus replication 16
Effects of virus infection on the host cell 37
Types of virus infection 37
Viral transformation and oncogenesis 41
Sub-viral infections 44
Further reading 50
Electronic resources 51

2 Viral interactions with the immune system **53**

The non-specific immune response 53
The cell-mediated immune response 55
The serological immune response 60
Apoptosis 67
The role of co-stimulation in the immune response 67
Evasion of immune surveillance 68
Epitope mapping 70
Production of specific antisera 73
Passive immunity 76
Further reading 78
Electronic resources 79

3 Vaccines and immunotherapy **81**

Current vaccines 81
Adjuvants 86

Approaches to vaccine development 90
Possible pathogenic effects of biotechnology-derived vaccines 94
Tailoring of the immune response to vaccination 95
Alternative delivery systems 96
Therapeutic vaccination 97
Further reading 98
Electronic resources 98

4 Antiviral drugs **99**
The history of antiviral drug development 99
Current antiviral drugs 99
Resistance to antiviral drugs 104
Combination therapy 107
Interferons 109
Development of antiviral drugs 110
Nucleic acid-based approaches to antiviral drug development 115
Further reading 118
Electronic resources 119

5 Cloning and gene therapy **121**
Viruses and cloning 121
Construction of a plasmid vector 121
Cloning of viral genes 123
Expression in eukaryotic cells 125
Viral vector systems 128
Gene therapy 129
Further reading 133
Electronic resources 133

6 Molecular diagnostics **135**
Immunological assays 135
Nucleic acid detection and amplification 137
Future developments in diagnostic virology 146
Further reading 148
Electronic resources 149

7 New and emerging viruses **151**
Where do new or emerging viruses come from? 151
What factors contribute to disease emergence? 167
Monitoring emerging infectious diseases 169
Further reading 170
Electronic resources 171

Appendix A. Glossary **173**

Index **179**

Abbreviations

AAV	adeno-associated virus
ADCC	antibody-dependent cellular cytotoxicity
AIDS	acquired immune deficiency syndrome
ARDS	acute respiratory distress syndrome
AZT	3′-azido-3′-deoxythymidine
BSE	bovine spongiform encephalopathy
CD	clusters of differentiation
CDC	Centers for Disease Control, Atlanta, Georgia
cDNA	DNA complementary to a messenger RNA, produced using reverse transcriptase
CDR	complementarity-determining region
CH	constant region of the immunoglobulin heavy chain
CJD	Creutzfeld–Jakob disease
CL	constant region of the immunoglobulin light chain
CMI	cell-mediated immune response
CMV	cytomegalovirus
CTL	cytotoxic T lymphocyte (T cell)
DI	defective interfering
DISC	disabled infectious single cycle
DNA	2′ deoxyribonucleic acid, a polymer of the monophosphate forms of deoxyadenosine (A), deoxycytidine (C), deoxyguanosine (G) and deoxythymidine (T)
ds	double-stranded
EBNA	Epstein–Barr (virus) nuclear antigen
EBV	Epstein–Barr virus
Fc	region of immunoglobulin molecule from the hinge region (or equivalent) to the carboxyterminus
gp	glycoprotein
GSS	Gerstmann–Sträussler (Scheinker) syndrome
HA	hemagglutinin
HAV	hepatitis A virus
HDV	hepatitis delta virus
HFRS	hemorrhagic fever with renal syndrome
HHV	human herpesvirus

HIV	human immunodeficiency virus
HPS	hantavirus pulmonary syndrome
HPV	human papilloma virus
HSV	herpes simplex virus
HTLV	human T-cell leukemia virus
ICAM	intercellular adhesion molecule
Ig	immunoglobulin
IL	interleukin
J-chain	joining protein of immunoglobulins M or A
kb	kilobase
kbp	kilobase pair
KS	Kaposi's sarcoma
KSV	Kaposi's sarcoma virus (properly, human herpesvirus 8)
LCMV	lymphocytic choriomeningitis virus
LCR	ligase chain reaction
LFA	lymphocyte function-associated antigen
LTR	long terminal repeat
MHC	major histocompatibility complex
MMR	measles/mumps/rubella
mRNA	messenger RNA
NA	neuraminidase
NANB	non-A, non-B
NASBA	nucleic acid sequence-based amplification
NGF	nerve growth factor
NK	natural killer
nvCJD	new variant Creutzfeld–Jakob disease
OPV	oral polio vaccine
PCR	polymerase chain reaction
PrP	prion protein
PSTV	potato spindle tuber viroid
Rb	retinoblastoma
REA	restriction endonuclease analysis
RF	replicative form
RFLP	restriction fragment length polymorphism
RI	replicative intermediate
RNA	ribonucleic acid, a polymer of the monophosphate forms of adenosine (A), cytidine (C), guanosine (G) and uridine (U)
RRE	Rev-response element
RSV	respiratory syncytial virus
RT	reverse transcriptase
SIV	simian immunodeficiency virus
ss	single-stranded
SSPE	sub-acute sclerosing pan-encephalitis

SV40	simian virus 40
TAP	transporter (associated with) antigen processing
TAR	*trans*-activation response element
TCR	T-cell receptor
TGF	transforming growth factor
TK	thymidine kinase
TNF	tumor necrosis factor
TSE	transmissible spongiform encephalopathy
UNG	uracil *N*-glycosylase
UV	ultraviolet
VH	variable region of immunoglobulin heavy chain
vhs	virion shut-off
VL	variable region of immunoglobulin light chain
VLP	virus-like particle
VP	viral protein
VSV	vesicular stomatitis virus
VZV	varicella-zoster virus

Preface

This book is intended to provide a solid grounding in biomedical virology, to provide the basis for understanding the applications of 'molecular virology' in the biomedical sciences. As always, in a book of this length, much has had to be left out, but I have taken advantage of the publication of the second edition to add sections on areas of particular importance as well as to revise and update the book as a whole.

David R. Harper

Acknowledgments

I should like to thank all those who provided material for inclusion in this book, including Drs Adams, Chrystie, Ebbs, Grose, Morrow, Ng, Nye and Parkin, as well as the authors of the *Sourcebook of Medical Illustration*. I should also like to thank my various proofreaders for their comments and advice, and Jonathan Ray and the staff at BIOS for their support.

This book is dedicated to Adam, Thomas, Peter, Christopher and Amanda.

Chapter 1

Virus structure and replication

1.1 What is a virus?

A virus is a subcellular organism with a parasitic intracellular life cycle. It has no metabolic activity outside the host cell, and contains either DNA or RNA, but not both, unlike cells. Outside the host cell, a virus is not actually alive. Rather, it has the potential for life, in the same way that a disk containing the code for a computer program is only a potential program until it is put into the host computer. Viruses are not 'the simplest form of life', since their life cycle involves not only their own metabolism, but also that of the cell whose replicative machinery they use. For all of that, viruses provide us with relatively straightforward approaches to studying biochemical events, and have helped to explain many aspects of biology. A generalized outline of virus replication is shown in *Figure 1.1*.

1.1.1 What makes a virus?

At its most basic, a virus particle (a 'virion') consists of a nucleic acid genome surrounded by a shell of protein, which may also contain lipids and sugars. The basic function of a virion is to deliver the viral genome into a cell where it can replicate. This requires:

(1) structures to contain and protect the nucleic acid genome together with any associated proteins which are required for its replication;
(2) specific receptors/effectors on the virion surface which will allow the virus to enter the target cell.

These requirements are fulfilled in very different ways by different viruses. The complexity of a virus is a direct reflection of the size of the viral genome. The more information a virus can encode, the more proteins it can make. Viruses infecting humans produce anything from one to more than a hundred proteins. Viruses come in a range of shapes and sizes ('morphologies'), summarized in *Figure 1.2*.

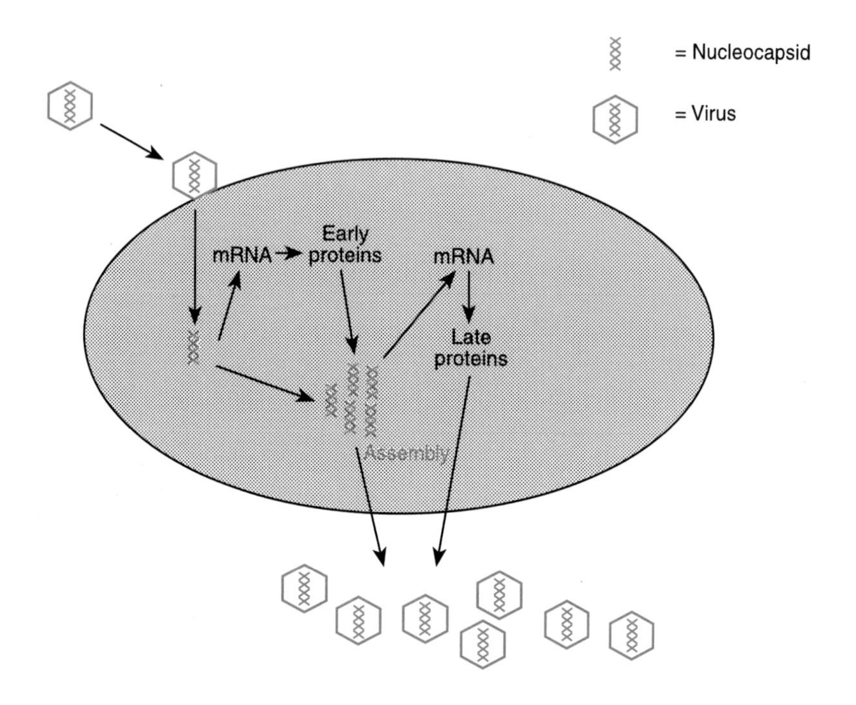

Figure 1.1: Schematic diagram of virus replication.

The most basic virus particle is made up of protein and nucleic acid. The proteins are the structural components and are also the effectors which are responsible for the infection of cells and the production of new ('progeny') viruses, while the nucleic acid provides the viral genetic code required to produce the proteins. Many viruses contain carbohydrate attached to glycosylated proteins of the viral surface. Lipid may also be present, derived from the membranes of the host cell, most commonly as an envelope around the outside of the virus, although some viruses do have smaller amounts of lipid within the virus particle.

The viral genome can be either DNA (as with all cellular life) or (uniquely to viruses) RNA. Virions do not contain both types of nucleic acid. An RNA genome, while relatively common among viruses, has many implications for the virus, discussed in Section 1.4.5. The genome is contained in a nucleoprotein structure, the nucleocapsid, complexed with the capsid proteins (structural proteins and replicative enzymes).

The viral envelope (where present) is a membrane, derived from the host cell but with viral proteins embedded in it, that surrounds the capsid.

1.2 Virus classification

The formal taxonomical system for the classification of viruses is administered by the International Committee for the Taxonomy of

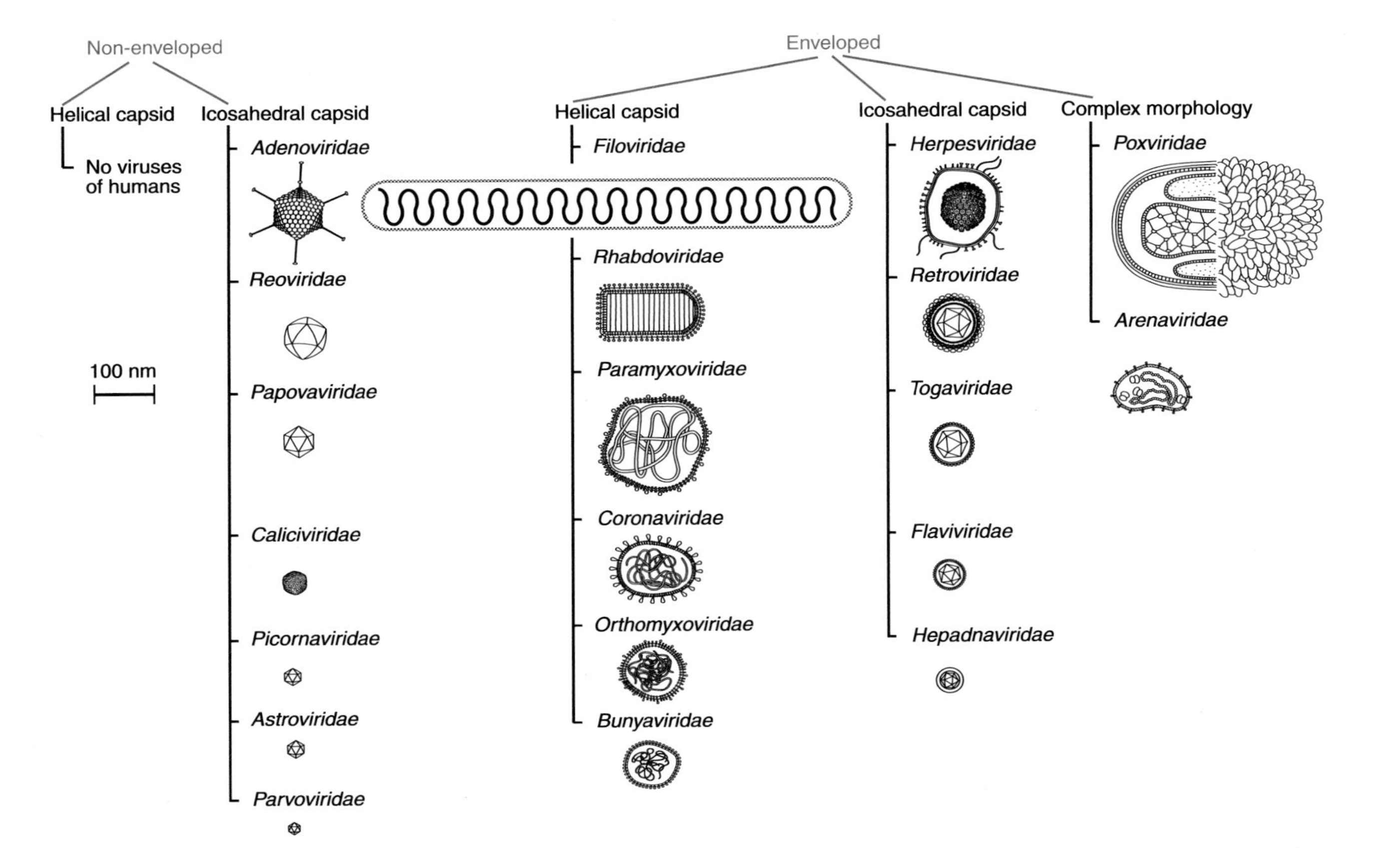

Figure 1.2: Viruses infecting humans. Diagrams reproduced from the *Sourcebook of Medical Illustration*. The appearances shown are those seen using 'classical' transmission electron microscopy; hydrated forms of viruses often have different appearances.

Table 1.1: Properties of the main virus families infecting humans

Virus family	Genome type (sense) and size	Morphology/ capsid symmetry	Size (nm)	Major virion proteins	Examples of human disease
Adenoviridae	dsDNA 30–38 kbp	Unenveloped/ icosahedral	70–90	⩾10	Common cold
Arenaviridae	ssRNA (±) 10–12 kb in two segments	Enveloped/ complex	110–130	2–3	Lassa fever
Astroviridae	ssRNA(+) 7–8 kb	Unenveloped/ icosahedral	27–34	1–6	Diarrhea
Bunyaviridae	ssRNA (±) 10–24 kb in three segments	Enveloped/ icosahedral	80–100	4	Rift valley fever, respiratory disease
Caliciviridae	ssRNA (+) 15 kb	Unenveloped/ icosahedral	35–40	2	Hepatitis E, Norwalk gastroenteritis
Coronaviridae	ssRNA (+) 27–33 kb	Enveloped/ helical	60–220	3–4	Common cold
Filoviridae	ssRNA (−) 13.5 kb	Enveloped/ helical	80 x 800–14 000	7	Hemorrhagic fevers
Flaviviridae	ssRNA (+) 12 kb	Enveloped/ icosahedral	40–60	3–4	Yellow fever
Hepadnaviridae	ss/dsDNA 3.2 kb (total) 1.7–2.8 kbp (ds)	Enveloped/ icosahedral	40–48	7	Hepatitis B
Herpesviridae	dsDNA 120–225 kbp	Enveloped/ icosahedral	100–200	> 20	Herpes simplex, chickenpox

Orthomyxoviridae	ssRNA (−) 13.5 kb in 7–8 segments	Enveloped/ helical	90–120	7	Influenza
Papovaviridae	dsDNA 4.5–7.5 kbp	Unenveloped/ icosahedral	40–55	6–9	Papillomas, progressive multifocal leukoencephalopathy
Paramyxoviridae	ssRNA (−) 15–21 kb	Enveloped/ helical	125–250	10–12	Measles, mumps, respiratory diseases
Parvoviridae	ssDNA (±) 4.5–6 kb	Unenveloped/ icosahedral	18–22	2–3	B19 infection (exanthem)
Picornaviridae	ssRNA (+) 7–9 kb	Unenveloped/ icosahedral	30	4	Common cold, polio
Poxviridae	dsDNA 130–375 kbp	Enveloped/ complex	220–450 × 140–260	> 100	(Smallpox) *Molluscum contagiosum*
Reoviridae	dsRNA 18–30 kbp in 10–12 segments	Unenveloped/ icosahedral	60–80	6–10	Diarrhea, Colorado tick fever
Retroviridae	ssRNA (+) dimer, each 7–10 kbp	Enveloped/ icosahedral	80–100	8	AIDS, human T-cell leukemias
Rhabdoviridae	ssRNA (−) 10–14 kb	Enveloped/ helical	100–430 × 45–100	5	Rabies
Togaviridae	ssRNA (+) 12 kb	Enveloped/ icosahedral	60–70	3-4	Rubella, encephalitis

Viruses, and is based on several structural properties:

- size (anything from 20 nm diameter sphere (*Parvoviridae*) to up to a 450×250 nm multi-layered brick-shaped mass (*Poxviridae*) (*Figure 1.2*); some viruses may produce even larger forms (*Filoviridae*);
- nucleocapsid symmetry (helical, icosahedral or complex);
- presence of an envelope membrane (originally defined as ether-resistant or -sensitive, since the organic solvent will destroy the envelope membrane);
- type of genome (RNA or DNA).

Genetic relationships determined from nucleic acid sequence data are now also used to determine or refine relationships between viruses.

Viruses are classified into families (ending *-viridae*), and genera (ending *-virus*) on the basis of these criteria. The classical taxonomical definition of a species is not used for viruses. Rather, viruses are identified by clinical, immunological, structural and molecular means and are then grouped into the categories defined above. Such groupings may change as more information about the viruses is obtained. Two other levels of classification may be used, orders (ending *-virales*, containing several virus families), and subfamilies (ending *-virinae*, between family and genus). Only two orders containing viruses infecting humans have been defined to date, the orders *Mononegavirales* (families *Filoviridae, Paramyxoviridae* and *Rhabdoviridae*) and *Nidovirales* (*Coronaviridae* and *Arteriviridae*). The recently defined family *Bornaviridae* has also been assigned to the *Mononegavirales*, but infection of humans is still controversial. The properties of virus families infecting humans are shown in *Table 1.1*.

Most viruses can be readily classified by these criteria, and they provide a straightforward basis for the taxonomical groupings used. Viruses are grouped by these criteria in *Figure 1.2*. Most viruses have a 'standard' morphology and fall into one of the main groups, defined as:

- unenveloped/helical capsid;
- unenveloped/icosahedral capsid;
- enveloped/helical capsid;
- enveloped/icosahedral capsid.

Examples of each of these types are shown in *Figure 1.3*.

1.2.1 Complex viruses

As with most systems, there are many exceptions, and some viruses have morphologies that differ from that of the main groups. Two of the more unusual examples of virus morphology (poxviruses and filoviruses) are discussed in more detail below. In addition, most viruses can produce forms which are totally unlike their 'classical' appearance – influenza is particularly variable when grown in tissue culture, as shown in *Figure 1.4* – and the common descriptions of virus morphology may be very different from what is actually seen under the electron microscope.

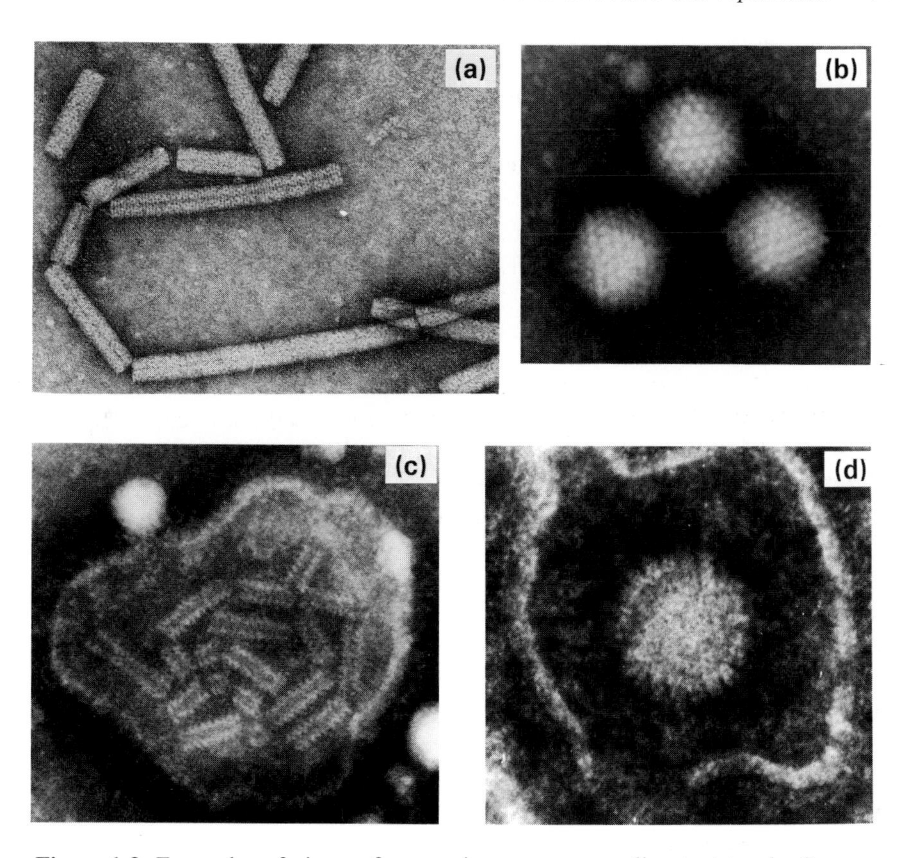

Figure 1.3: Examples of viruses from main groups according to 'standard' morphology. (a) Unenveloped/helical (tobacco mosaic virus), (b) unenveloped/icosahedral (adenovirus), (c) enveloped/helical (paramyxovirus), (d) enveloped/icosahedral (herpesvirus). Photographs courtesy of Dr Ian Chrystie, Department of Virology, St Thomas' Hospital, London, and Professor C.R. Madeley, Department of Virology, Royal Victoria Infirmary, Newcastle-upon-Tyne.

Poxviruses (including smallpox and vaccinia) have the largest known viral genomes and are the most complex of viruses, producing over a hundred proteins. Poxviruses are just about visible by light microscopy, and can be almost 0.5 μm long. A poxvirus is often described as 'brick-shaped', and is in some ways reminiscent of a cell-like structure. It has an outer coat containing lipid and protein structures and complex internal structures including a 'core' containing the genome. The outer lipid–protein shell is not the envelope commonly seen with other viruses, but rather is made up of lipoprotein subunits which form the virion surface. In the electron microscope the two typical appearances are referred to as the C (capsule) and M (mulberry) forms: the former is smooth and featureless, while in the latter the repeating subunits can be seen (*Figure 1.5*). Inside this is a complex arrangement of proteins, containing the many viral enzymes that a poxvirus uses to replicate. The M form, while often shown

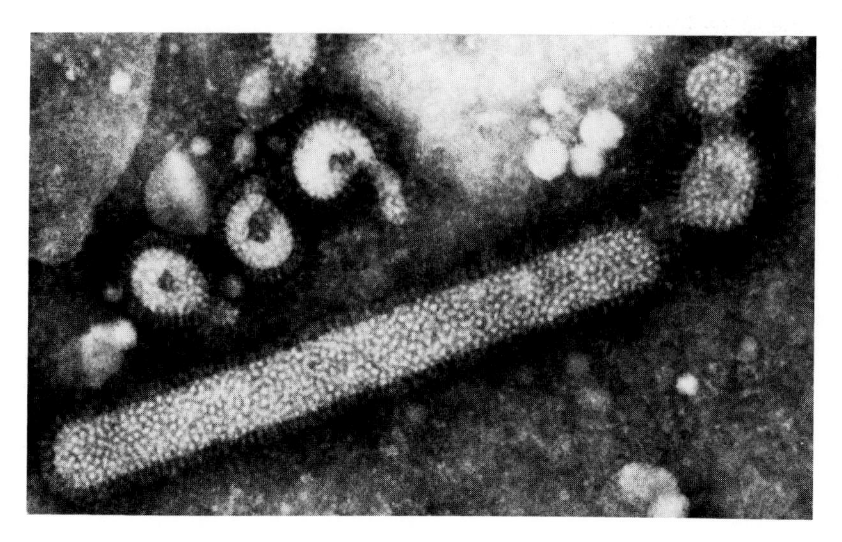

Figure 1.4: Variable forms of influenza virus. Photograph courtesy of Dr Ian Chrystie, Department of Virology, St Thomas' Hospital, London.

in textbooks, is almost certainly an artifact resulting from drying during preparation for electron microscopy.

Filoviruses (Marburg and Ebola hemorrhagic fever viruses) have an approximately helical nucleocapsid in an extremely long enveloped particle which, although only 80 nm wide, can be up to 14 000 nm long, making them easily the largest viruses. However, these are very simple viruses (with a genome about one-twentieth the size of that of the poxviruses), and their massive length appears to be the result of poorly controlled elongation rather than a genuinely complex structure. In support of this, maximum infectivity is actually associated with the shorter (but still very large) 800–900 nm forms (*Figure 1.5*).

Many viruses contain additional elements. Examples include the thick proteinaceous tegument between the herpesvirus envelope and nucleocapsid or the cellular ribosomes which are incorporated into arenaviruses.

While it is usually relatively simple to identify the type of a virus by electron microscopy, the limitations of using virus structure are apparent with the small unenveloped icosahedral viruses. The icosahedral structure of these is usually unclear, and they are generically referred to as 'small round viruses', which is entirely reasonable since all that can be said from their morphology in the electron microscope is that they are small, round and (probably) viruses.

1.3 Virus structure

1.3.1 Capsids

The capsid is the protein structure surrounding the viral genome and is formed of repeating protein subunits ('capsomers') assembled around the

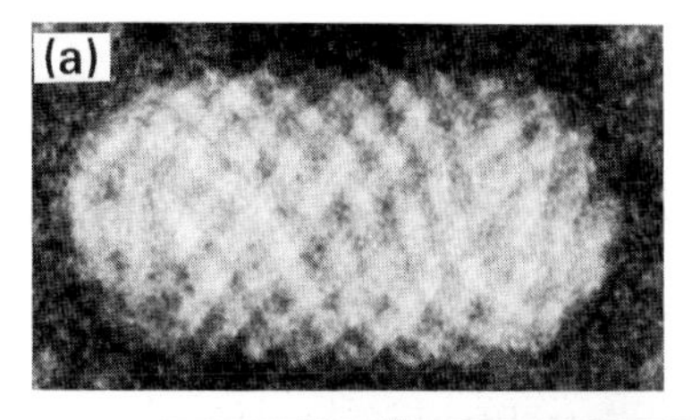

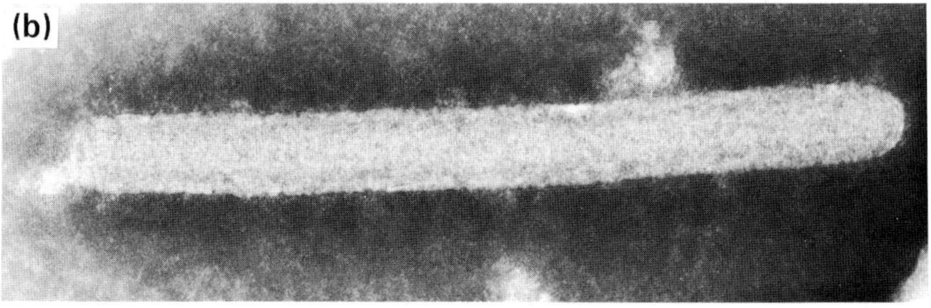

Figure 1.5: Forms of poxvirus (a) and filovirus (b). Photographs courtesy of Dr Ian Chrystie, Department of Virology, St Thomas' Hospital, London, and Dr A.M. Field, Central Public Health Laboratory, Colindale.

nucleic acid, the whole being referred to as the nucleocapsid. The use of small protein subunits reduces the amount of genetic coding capacity that has to be dedicated to producing the capsid proteins, which is important since even the largest viral genomes are very small by cellular standards (*Figure 1.6*). Since protein components will naturally align in the most energetically favorable state, certain structures are favored. These most commonly have helical or icosahedral symmetry. In a capsid with helical symmetry, proteins are aligned in a helix around the nucleic acid and are rod-like in appearance. An icosahedron is a 20-sided solid with faces formed of identical equilateral triangles. Some viruses (e.g. adenovirus, *Figure 1.3b*) are visibly of icosahedral shape, but while capsids with icosahedral symmetry are all roughly spherical, most are not clearly icosahedral – symmetry is not the same as shape. However, all such capsids have the two-, three- and fivefold axes of rotational symmetry around the face edge (2), face center (3) or face vertex (5) which are what define icosahedral symmetry (*Figure 1.7*). The subunits making up the capsids of tobacco mosaic virus and adenovirus are also shown in *Figure 1.7*, and illustrate the use of repeating subunits in helical and icosahedral capsids, respectively. The complex rules of geometry which govern the precise formation of capsids are give in detail elsewhere and lie outside the scope of this book. Interested readers are referred to the list of further reading at the end of this chapter.

1.3.2 Virus envelopes

Another important structural element present in many viruses is the envelope. This is derived from the membranes of the host cell, although

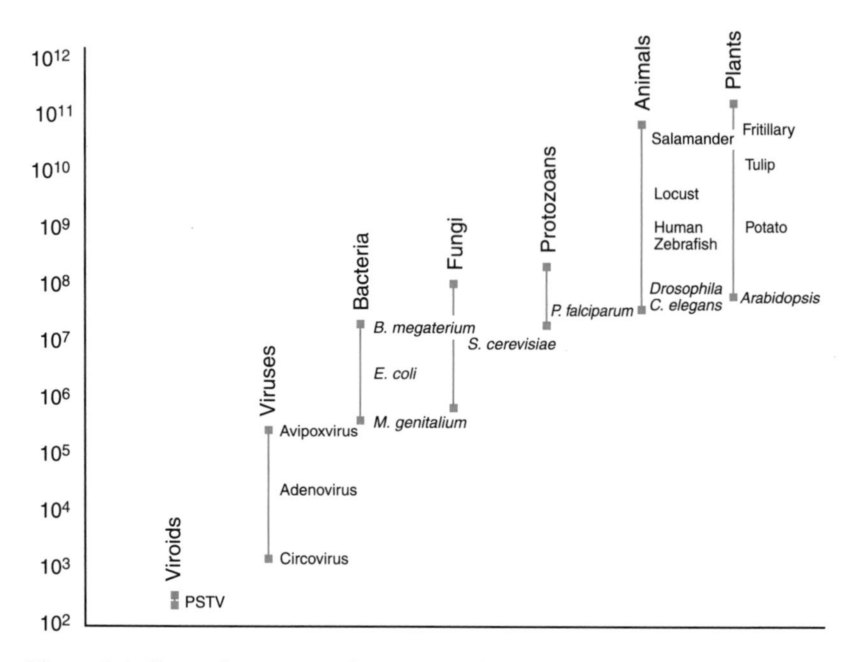

Figure 1.6: Sizes of genomes of representative organisms (kilobases/kilobase pairs). PSTV, potato spindle tuber viroid.

precisely which cellular membrane varies between viruses. Envelopment is not simply a passive process of picking up some membrane from the cell, since specific changes do occur to the membrane before it envelops the virus. Changes to membrane fluidity resulting from the preferential incorporation of specific lipids may be important. However, the most apparent change is the presence of viral proteins (seen as 'spikes' or a 'fringe' when the virion is viewed by electron microscopy, see *Figure 1.4*) projecting through the envelope. Clearly, viral proteins must be present on the outside of the envelope membrane in order to perform specifically viral functions such as binding to the host cell. These proteins are usually glycoproteins, with sugar groups attached to the polypeptide. The sugars make the protein locally hydrophilic, and are usually essential for function. All enveloped viruses have such proteins and, due to their nature and their position on the surface of the virion, they are usually highly immunogenic.

1.3.3 Viral genome size and its effect on virus structure

Having reviewed the basics of virus structure, it is now worth noting one of the major elements in the complexity of a virus. A large viral genome, while still very small by cellular standards (see *Figure 1.6*), means that the virus can produce many proteins, allowing a complex structure. Viruses with small genomes are more restricted. In general, the bigger the viral genome, the more complex the structure can be. An excellent example is

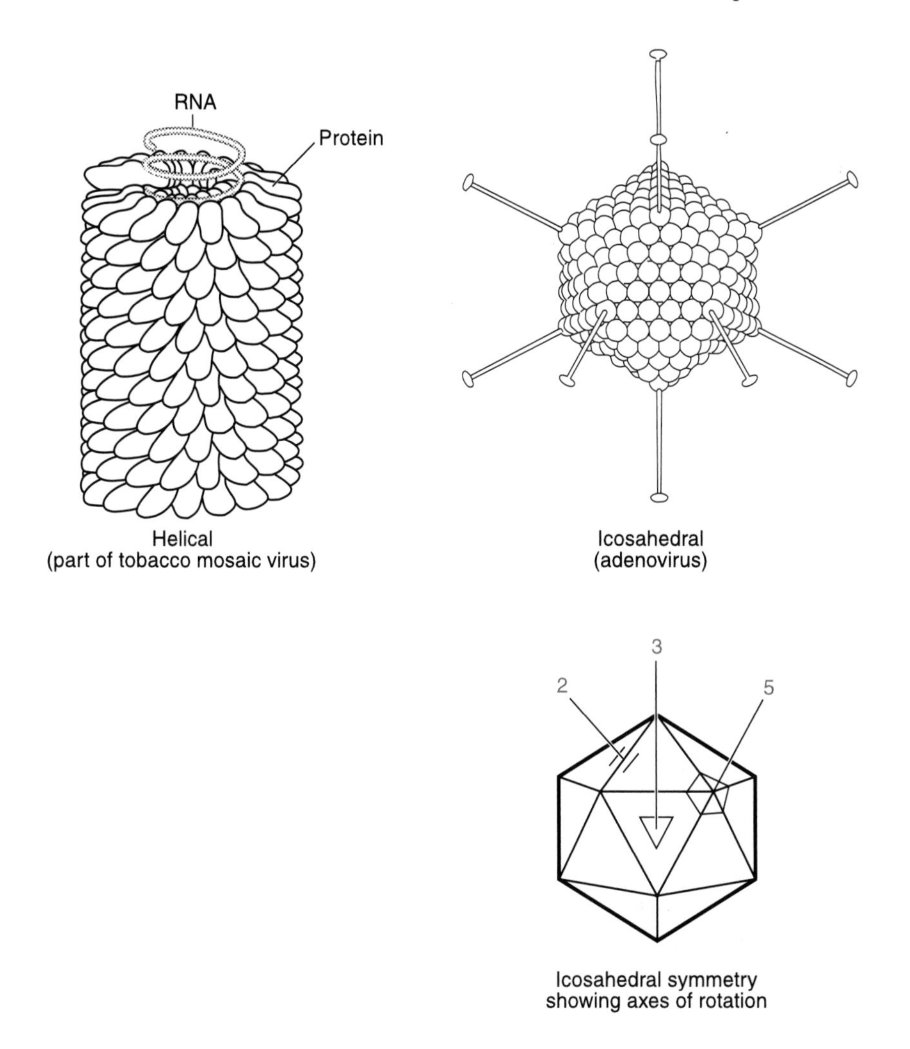

Figure 1.7: Helical and icosahedral capsid structures.

provided by comparing adenoviruses with picornaviruses. The adenovirus genome is a double-stranded DNA of 30–38 kilobase pairs (kbp), compared with the 7.2–8.5 kilobase (kb) single-stranded RNA genome of a picornavirus. Both virions are non-enveloped with icosahedral capsid symmetry. However, the fourfold difference in genome size allows the adenovirus particle to have a far more complex structure. The picornavirus particle is made up of only four main viral proteins (VPs 1–4), plus one copy of a very small protein (VPg) on the end of the genomic RNA. All are cleaved from a single precursor protein. VP1, VP2 and VP3 form the outer face, underlaid by the VP4 protein. Each face of the capsid is made up of five triangular subunits, each containing one copy of each protein (*Figure 1.8*). These pentamers then form the 12 faces of a dodecahedron. Nevertheless, this structure has the two-, three- and

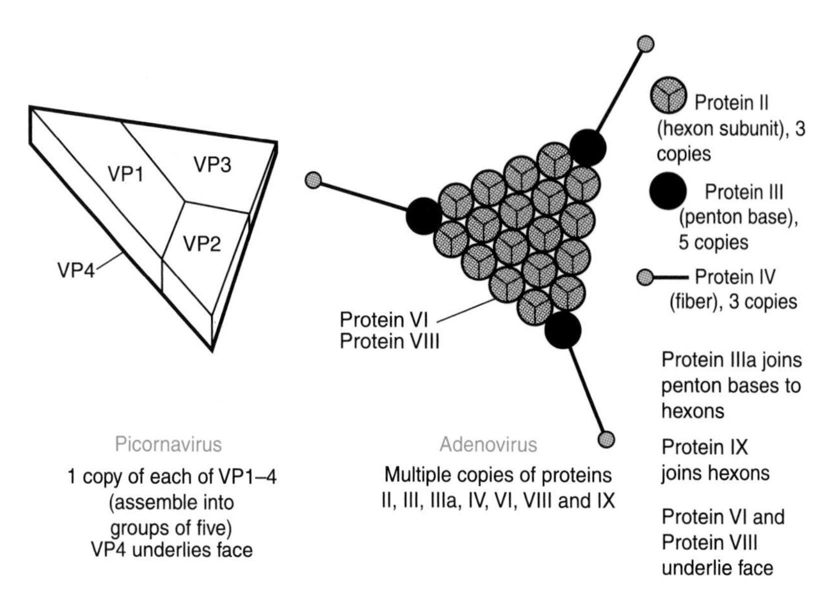

Figure 1.8: The capsid faces of icosahedral viruses.

fivefold axes of icosahedral symmetry. By contrast, adenovirus capsids are formed of 'hexons', each of which contains three copies of protein II. These hexons are joined by protein IX and are underlaid by proteins VI and VIII, while penton (vertex) faces contain five copies of protein III to which protruding fibers made of three copies of protein IV are attached (*Figure 1.8*). Penton bases are held to the hexons by protein IIIa. The hexons and pentons together make up the triangular faces of a regular icosahedron (*Figure 1.7*). The adenovirus particle as a whole contains at least 10 proteins. Despite these differences, the basic structure achieved has icosahedral symmetry in both cases. Even simpler than picornaviruses are the parvoviruses, which have the smallest genomes of any virus infecting humans. The parvovirus virion is 18–22 nm in diameter with icosahedral symmetry, and contains only three different types of protein. This contrasts with the poxviruses, which have the largest viral genomes (about 50 times the size of the parvovirus genome), and have highly complex virions containing more than 100 different proteins in multiple structures with an intricate 'cell-like' appearance (*Figure 1.9*).

In addition to the implications of a large genome for the complexity of virus structure, viruses with larger genomes are able to provide more of the functions that they need to replicate, and therefore they rely less on the production of specific enzymes by the cell. This can be useful to the virus, since many cells may not make all of the enzymes required by a particular virus unless they are actively dividing or unless they are specialized in some way.

With small virus genomes, a single protein frequently has multiple functions, since there is simply not enough coding capacity to produce one

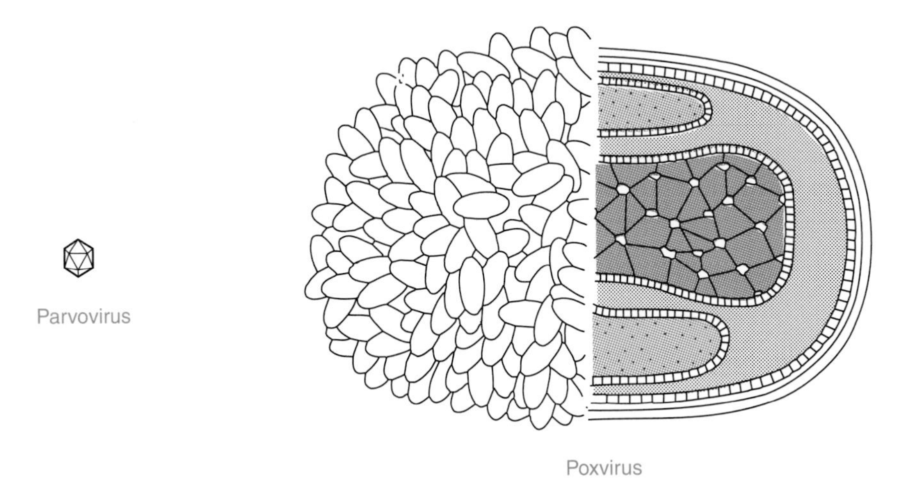

Figure 1.9: Structures of parvovirus and poxvirus. Reproduced from the *Sourcebook of Medical Illustration*.

protein for each task. The capsid proteins of the picornaviruses or papovaviruses, for example, can assemble themselves into capsids. This ability to 'self-assemble' is common to many viruses with small genomes. By contrast, the proteins of the herpesvirus and adenovirus capsids are not capable of assembling into the final structure without the help of other proteins which do not form part of the final virion. As stated above, capsid assembly involves the building of the capsid from small repeating protein subunits. This has the effect of minimizing the need for dedicated genome space to produce the structural proteins, an important consideration when even the largest virus has a genome one tenth the size of that of a typical bacterium, and a typical virus genome is smaller by a factor of 200 (*Figure 1.6*).

Due to the small sizes of their genomes, many viruses use methods of making one genome sequence produce multiple proteins. These methods are summarized in *Figure 1.10*. However, the use of such mechanisms is not restricted to viruses with small genomes. Adenoviruses (30–38 kbp) cut and join ('splice') mRNAs differently to produce different proteins from one original RNA, while the RNAs produced by latent Epstein–Barr or herpes simplex herpesviruses from genomes of greater than 150 kbp are also spliced. Herpesviruses also use multiple start sites for translation of mRNAs to produce some proteins, while retroviruses use inefficient stop sites to produce low levels of proteins from the mRNA after such sites.

An analogous system is used at a transcriptional level by paramyxoviruses with inefficient transcription stop sites between genes (see Section 1.4.5).

Since each amino acid is encoded by three bases (a codon), the protein produced by reading the mRNA can be produced in one of three 'reading frames', and can be quite different when produced from a different reading

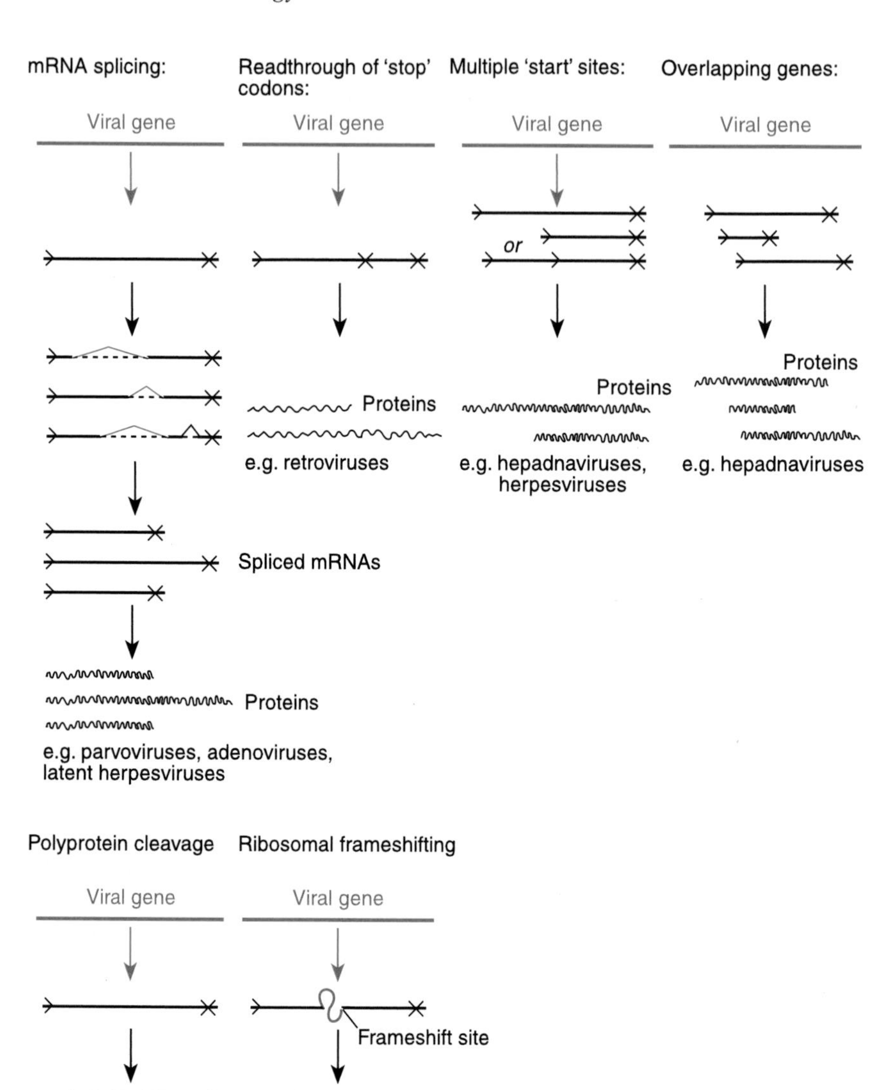

Figure 1.10: Strategies for making multiple proteins from one gene.

frame (*Figure 1.11*). Thus allows very different proteins to be produced from 'overlapping genes' in one nucleic acid sequence, but requires that the sequence 'make sense' when read two or three different ways. The use of different reading frames can also result from 'slippage' of the

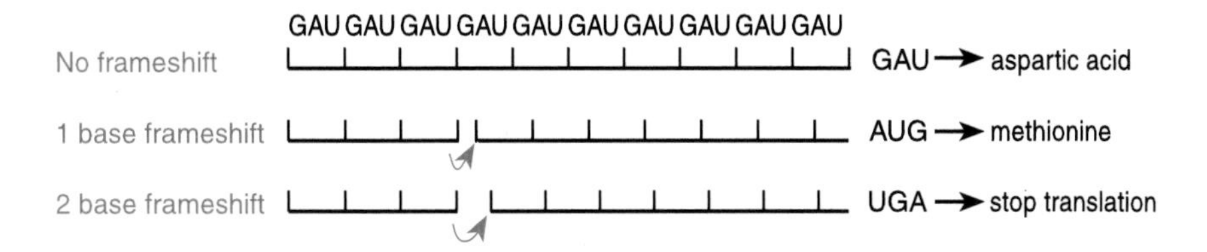

Figure 1.11: Effects of frameshifting and the use of different reading frames. A one-base frameshift alters translation from 'reading frame' 1 (aspartic acid) to reading frame 2 (methionine); a two-base frameshift alters translation from 'reading frame' 1 (aspartic acid) to reading frame 3 (stop codon); a three-base frameshift resumes translation in the original 'reading frame'.

translating ribosome at a frameshift site. In retroviruses, this site is a combination of a sequence at the gene junction and a fold in the mRNA downstream of this site.

Polyprotein cleavage can produce different proteins by cutting the polyprotein at different sites.

These methods are used by viruses, but not exclusively. For example, most eukaryotic genes appear to contain many introns (regions of RNA removed by splicing and not present in the mature mRNA), so it appears that size alone is not the reason for such strategies.

1.3.4 Viral genome type

The replication strategy of a virus is heavily influenced by the nature of the viral genome, and this is commonly used when classifying viruses. Indeed, a system which had quite wide acceptance was developed based entirely on this. Viruses may be subdivided by genome type, into those with:

- double-stranded (ds) DNA genomes (including the *Poxviridae, Herpesviridae, Adenoviridae* and the *Papovaviridae*). These are often among the largest of viral genomes;
- single-stranded (ss) DNA genomes (including the *Parvoviridae* and the *Circoviridae*, although the latter do not appear to infect humans). These are typically small genomes;
- dsRNA genomes (including the *Reoviridae* and the *Birnaviridae*, although infection of humans by *Birnaviridae* is not well characterized). As with all RNA genomes, these tend to be smaller than most DNA genomes. All classified viruses with dsRNA genomes have a genome consisting of between two and 12 different molecules of RNA (a 'segmented genome');
- ssRNA genomes, which may be subdivided into those which can function as mRNA (positive sense), and those which are complementary to the mRNA produced from them (negative sense). Viruses with positive sense RNA genomes include the *Picornaviridae* (a very large family with many members), *Caliciviridae, Coronaviridae, Flaviviridae* and *Togaviridae*. Viruses with negative sense RNA genomes include the *Orthomyxoviridae,*

Paramyxoviridae, Rhabdoviridae and *Filoviridae*. Most viruses of this type have a genome consisting of a single RNA molecule, but the *Orthomyxoviridae* have segmented genomes of 7–8 RNA molecules;
- viruses with RNA genomes that use a DNA intermediate stage (a provirus) to produce the RNA genome (*Retroviridae*);
- viruses with DNA genomes that use an RNA intermediate stage to produce the DNA genome (*Hepadnaviridae*).

The general processes of replication of viruses with the types of genome listed in this section are shown in *Figure 1.12*. However, before a virus can initiate replication, it must get into the cell.

1.4 Virus replication

1.4.1 Virus receptors

As stated earlier, a virion must protect the viral genetic material, and also be capable of delivering it into the host cell. The surface of any virus must have receptors which can bind to the target cell. The binding is mediated on one side by specific viral (glyco) proteins, and on the other side by cellular structures. Only a limited number of cellular receptors used by viruses are known (*Table 1.2*). The CD4 molecule provides a very good example of 'cell tropism', the targeting of a specific type of cell by the viral receptor. CD4 is a surface protein present only on the subset of T lymphocytes which are the main target cells for both the human immunodeficiency virus (HIV), which also requires other receptors as shown in *Table 1.2*, and human herpesvirus 7 (HHV-7).

1.4.2 Virus entry

Once the virus is bound to the cell, the next step is for it to get inside. Alternative methods of viral entry are shown in *Figure 1.13*. Many non-enveloped viruses are internalized by the cell itself, by translocation across the membrane or by the membrane invaginating to form a vacuole ('endocytosis'). These vacuoles frequently are acidified by 'proton pumps' in the membrane or by fusion with digestive endosomes (vacuoles containing acids and digestive enzymes) within the cell, in an effort to digest their contents. However, many viruses have adapted to use this by having proteins which, when their conformation is altered by the low pH, become able to mediate the exit of the viral genome (and other necessary components) into the cell itself. Examples include reoviruses and members of the *Adenoviridae*. However, it is thought that some members of the *Picornaviridae* use hydrophobic regions of their outer proteins to interact with the cellular membrane and thus allow the genome itself to penetrate into the cell (see Section 1.4.5). Some enveloped viruses [influenza (*Orthomyxoviridae*), Semliki Forest (*Togaviridae*) virus and members of the *Rhabdoviridae* and *Retroviridae*]

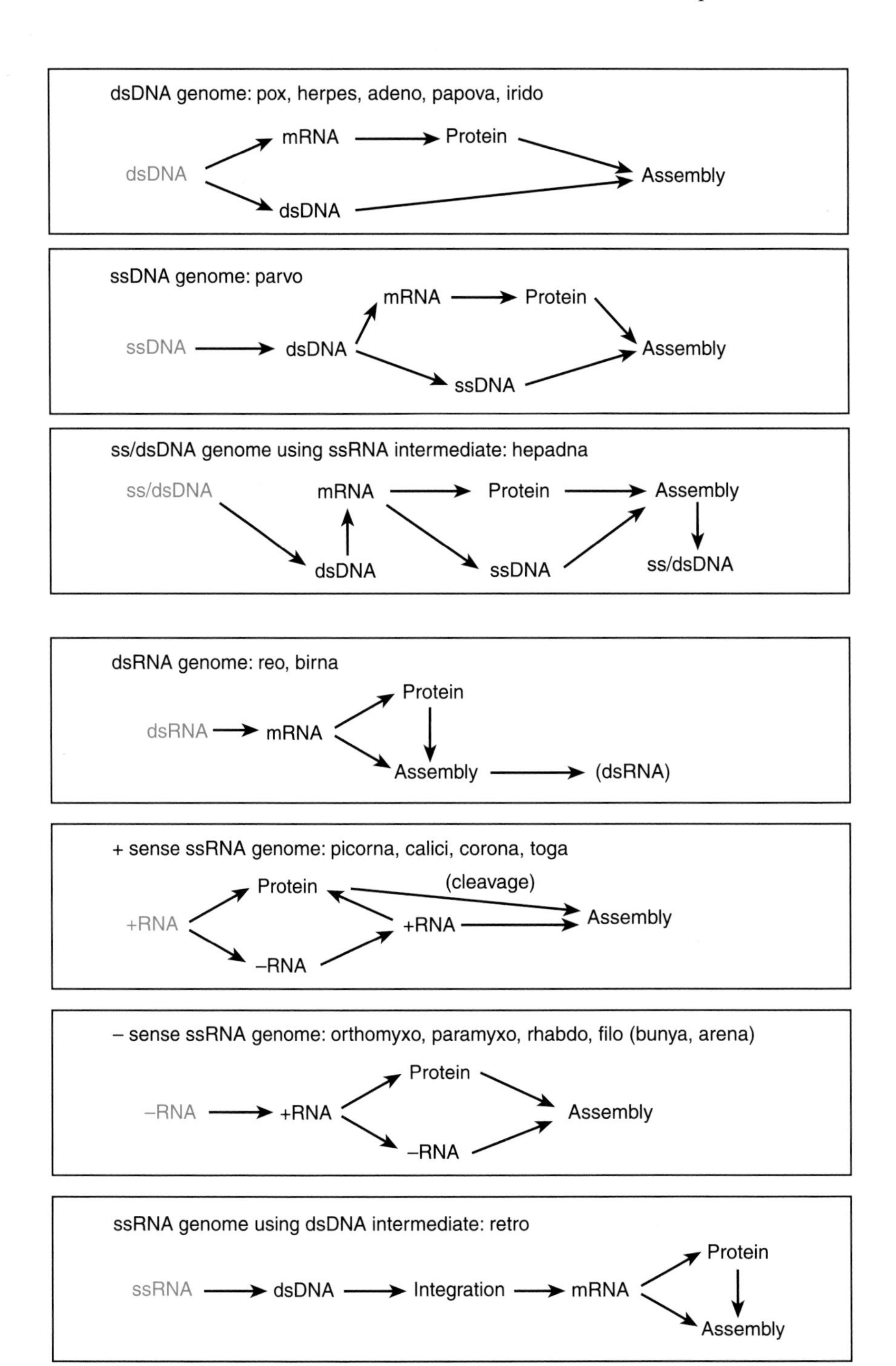

Figure 1.12: General methods of viral replication.

Table 1.2: Examples of receptors for viruses infecting humans

Family	Virus	Cellular receptor
Adenoviridae	Adenovirus type 2	Integrins $\alpha_v\beta_3$ and $\alpha_v\beta_5$
Coronaviridae	Human coronavirus 229E	Aminopeptidase N
Coronaviridae	Human coronavirus OC43	*N*-acetyl-9-*O*-acetylneuraminic acid on 120- or 30-kDa membrane proteins
Hepadnaviridae	Hepatitis B virus	IgA receptor, 35-kDa preS1-binding protein
Herpesviridae	Herpes simplex virus	Heparan sulfate proteoglycan plus mannose-6-phosphate receptor, or a TNF/NGF-related protein (HVEM) in some cell types
Herpesviridae	Varicella-zoster virus	Heparan sulfate proteoglycan plus mannose-6-phosphate receptor
Herpesviridae	Cytomegalovirus	Heparan sulfate proteoglycan plus second receptor
Herpesviridae	Epstein–Barr virus	CD21 (CR2) complement receptor
Herpesviridae	Human herpesvirus 7	CD4 (T4) T-cell marker glycoprotein
Orthomyxoviridae	Influenza A virus	Neu-5-Ac (neuraminic acid) on glycosyl group of membrane protein
Orthomyxoviridae	Influenza B virus	Neu-5-Ac (neuraminic acid) on glycosyl group of membrane protein
Orthomyxoviridae	Influenza C virus	*N*-acetyl-9-*O*-acetylneuraminic acid group on membrane protein
Papovaviridae	Papillomavirus	Alpha6 integrin complex
Paramyxoviridae	Measles virus	CD46 (MCP) complement regulator complexed with moesin
Parvoviridae	Adeno-associated virus 2 B19	150-kDa membrane glycoprotein Erythrocyte P globoside antigen
Picornaviridae	Coxsackieviruses B1, B3, B5, Echovirus 7, Enterovirus 70	CD55 decay-accelerating factor
Picornaviridae	Echoviruses 1 and 8	Integrin VLA-2 ($\alpha_2\beta_1$)
Picornaviridae	Foot-and-mouth disease virus	Heparan sulfate, integrin [α(v)β_3]
Picornaviridae	Hepatitis A virus	Mucin-like membrane glycoprotein (HAVcr-1)
Picornaviridae	Poliovirus	Immunoglobulin superfamily protein
Picornaviridae	Rhinoviruses	ICAM-1 adhesion molecule
Poxviridae	Vaccinia	Epidermal growth factor receptor
Reoviridae	Reovirus serotype 3	Sialic acid, β adrenergic receptor
	Rotavirus SA11	Sialic acid, gangliosides
Retroviridae	Human immunodeficiency virus	CD4 (T4) T-cell marker glycoprotein plus chemokine receptor (CC-CKR5, CXCR-4/fusin)
Rhabdoviridae	Rabies	Acetylcholine receptor, gangliosides

are internalized in vacuoles and then fuse with the vacuole wall as a result of the effect of the acidic environment on their outer proteins. The combination of a membrane and specific viral protein functions allows others actually to fuse with the cell surface and release their nucleocapsid directly into the cytoplasm. These include the *Paramyxoviridae* and *Herpesviridae* and some members of the *Retroviridae* (including HIV).

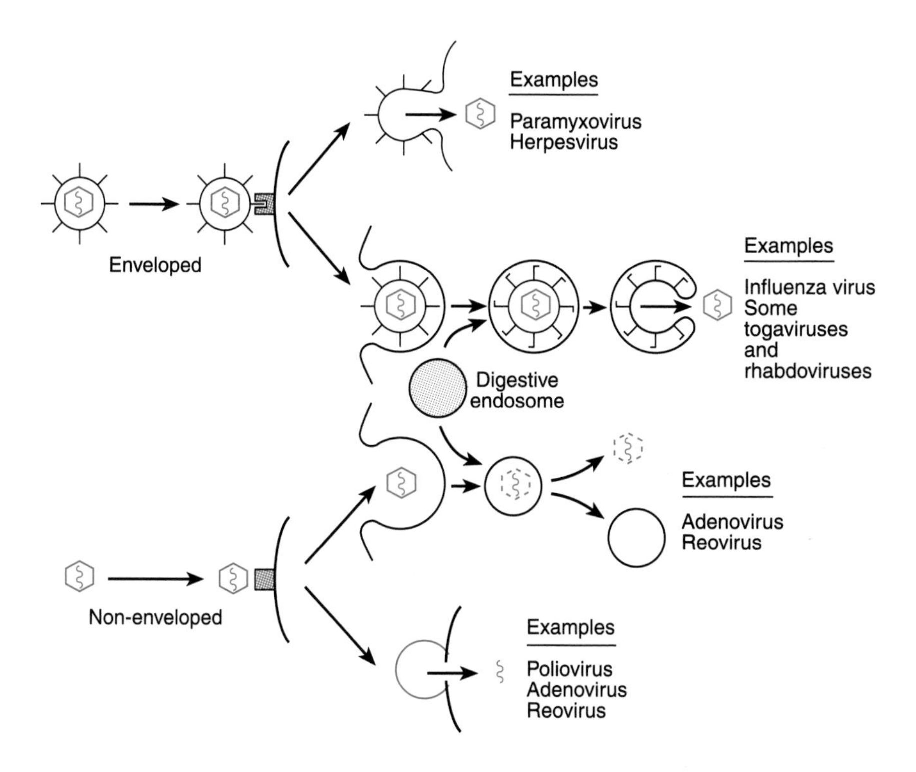

Figure 1.13: Methods of virus entry.

The end result of viral entry is that the viral nucleic acid–protein complex is within the cell, ready to begin making the components of the next generation of viruses. How this happens depends on the size and nature of the viral genome.

1.4.3 Virus replication

Where virus entry into the cell is followed by an active infection, this follows a 'one-step' growth curve (*Figure 1.14*). The first event is an 'eclipse phase', during which infectious virus cannot be recovered. In a eukaryotic cell, this lasts for anything from a few hours to a few days (but is much shorter in viruses of bacteria (bacteriophages), where it was first studied). The eclipse phase is the time when the virus has fragmented in order to begin replication but has not yet made and assembled the components of progeny virus. Following this stage, during an active (acute) virus infection, progeny virus begins to appear. The progeny virus genomes assemble into new viruses, and are also used to make yet more virus, and, of course, viruses produced from these progeny viruses may also be used, until the cell is totally dedicated to the production of viruses. This amplification results in a logarithmic increase in the amount of virus

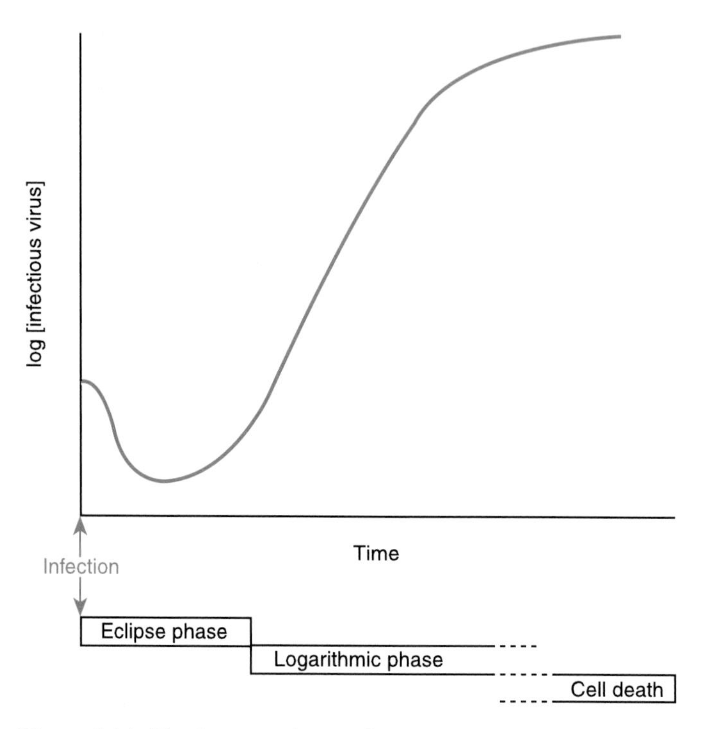

Figure 1.14: The 'one-step' growth curve.

present within the cell, and is referred to as secondary transcription. In an active infection, cell death then follows.

Owing to the limitations of space in a book of this size, it is not possible to detail all of the mechanisms used by viruses. Instead, a limited number of specific examples are presented.

1.4.4 DNA virus replication

Viruses with dsDNA genomes. These often have large genomes (up to 375 kbp), allowing them to have complex virions and to produce a range of virus-specific enzymes, unlike more limited viruses which depend to a very great extent on cellular enzymes and processes. For example, the enzyme thymidine kinase, which supplies components of DNA synthesis and is produced by *Herpesviridae* and *Poxviridae*, is only present in actively dividing cells. By producing viral versions of cellular enzymes, the virus ensures that they are available when and where the virus wants them. Poxviruses have a very large coding capacity, and set up what has been called a virus-coded 'second nucleus' in the cytoplasm, where replication occurs. While other complex DNA viruses such as the herpesviruses produce many enzymes with functions similar to cellular enzymes, their replication occurs in the nucleus.

The large DNA viruses also have distinct phases in their replication. Instead of producing all of the viral proteins at one time, they produce different groups of proteins at particular and appropriate times, often dependent upon prior viral syntheses and on specific viral control functions. Such timing is often controlled by the use of different 'promoters' (regions of the genome which promote transcription) for each group of proteins, which are activated by different but specific stimuli. One of the best studied examples of this is herpes simplex virus (*Figure 1.15*). After the virus enters the cell, the proteins contained in the viral tegument and capsid alter cellular functions and allow the synthesis of the first batch of viral mRNAs and proteins. These are referred to as 'immediate–early' or 'alpha' proteins, and are mainly regulatory proteins. These, in turn, are necessary to allow the synthesis of the next set of proteins, 'early' or 'beta' proteins. Most such proteins are enzymes concerned with synthesizing the viral DNA and also with preparing the cell for the manufacture of the virus. Early proteins, together with the newly synthesized viral DNA, allow the synthesis of the 'late' or 'gamma' proteins. These are mostly the structural components of the new virus. Most of the proteins that entered the cell with the virus were late proteins

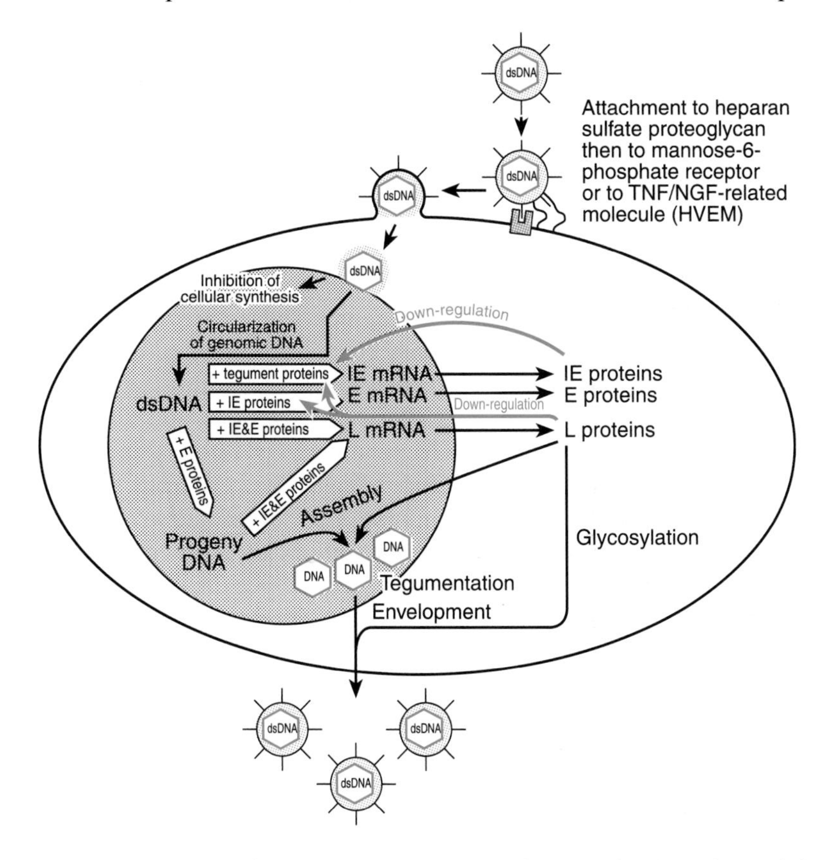

Figure 1.15: Herpes simplex virus replication (dsDNA virus). In latent infection, late mRNAs and proteins are not produced, and circularized dsDNA is maintained as an episome with very limited transcription.

(with some important exceptions). This completes the cycle by starting off the synthesis of the immediate–early proteins when the progeny virus infects a new cell.

DNA synthesis requires short (usually RNA) primers from which to start, and the nucleic acid to which these bind must also be replicated. The mechanisms by which full-length copies of any viral genome are made without missing out these, together with other regions at the ends of the molecule(s), are highly varied and often involve complex folding of the genome during transcription. The basic mechanism of replication for herpesviruses involves circularization of the (linear) herpesvirus genome. Viral dsDNA genomes are then 'rolled off' as very long polymers of the genome, known as 'concatamers', which are cut to unit length during viral assembly. The production of concatameric copies of the viral genome is very common, and is one (relatively simple) system to allow full-length genome copies to be produced, since the polymerase 'completing the circle' will displace the primers that initiated nucleic acid synthesis.

Herpesvirus replication takes place in the nucleus. This is quite common with DNA viruses, whereas many RNA viruses replicate in the cytoplasm. This reflects the fact that the production of DNA or RNA from a DNA template can use cellular polymerase enzymes which are present in the nucleus. However, cells do not routinely make RNA from an RNA template, so that a virus with an RNA genome is required to make and use viral enzymes. With the herpesviruses, the viral mRNAs must go to the cytoplasm to be translated to proteins, and then return to form capsids around the herpesvirus genomes. Herpesvirus capsids are assembled in the nucleus and then bud through the nuclear membrane before being transported to the cell surface in cellular vacuoles.

Smaller dsDNA viruses which cannot encode so many proteins (such as the *Papovaviridae*) use more cellular functions, and appear to have less complicated controls of replication. Papovaviruses use an alternative approach to obtaining enzymes only produced by dividing cells, and produce factors which induce cell division. This allows the virus to obtain the enzymes required for replication, and the forced switch into cell division is likely to be involved in the induction of cancer by these viruses.

Viruses with ssDNA genomes. All the viruses of this type have very small genomes, the largest being a bacterial virus genome of about 8.5 kb. The smallest is that of porcine circovirus, which at 1759 nucleotides is also the smallest genome of a virus known to infect vertebrates. The best studied examples of ssDNA viruses which infect vertebrates are members of the *Parvoviridae* (*Figure 1.16*). Their genome is a typical ssDNA viral genome of 4.5–6 kb. Once the virus is within the cell, the genome is copied in the nucleus to produce a dsDNA 'replicative form' (RF) (*Figure 1.17*). This is then used to produce both mRNA (and viral proteins) and progeny genomes. However, cells which are not actively dividing do not make DNA. Parvoviruses are among the most limited of viruses due to the small

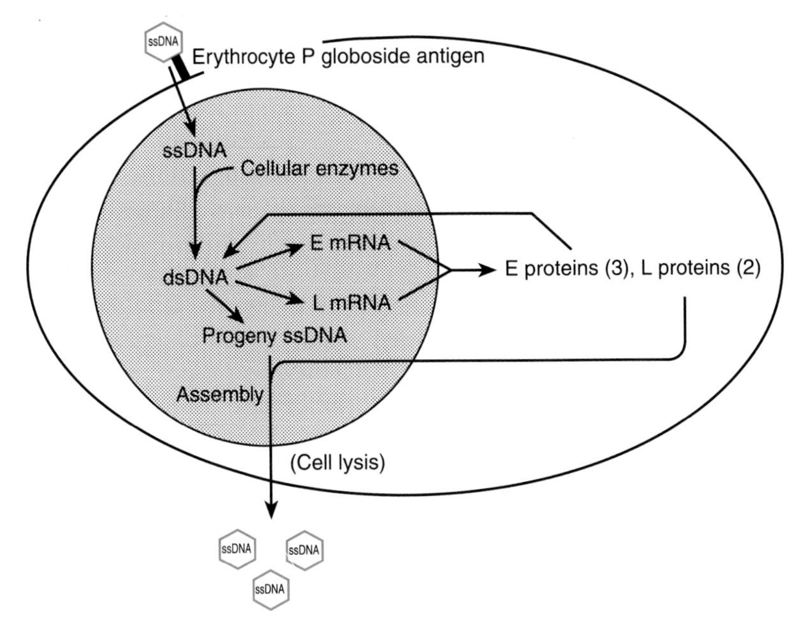

Figure 1.16: Parvovirus B19 replication.

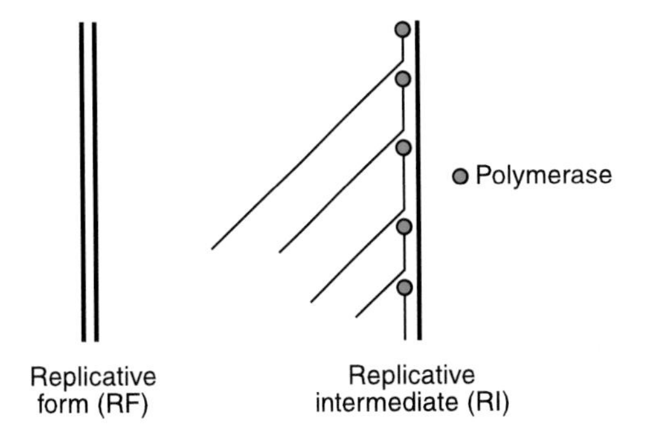

Figure 1.17: Replicative nucleic acid structures.

size of their genome, and do not have the ability to turn on cellular DNA synthesis. Instead, they can only infect productively a cell that is actively making DNA. This restricts the types of cell which will support their replication to actively dividing cells, such as erythrocyte precursor cells. Typically, parvoviruses make mRNAs for two to three structural proteins slightly later than mRNAs for one to three non-structural proteins. This means that there must be some rather limited form of temporal control, but the methods of this control are not clear and such control is not apparent at the level of protein synthesis.

dsDNA viruses that use an RNA intermediate. The only virus infecting humans that replicates in this way is hepatitis B virus, a member of the

Hepadnaviridae (*Figure 1.18*), although a similar strategy is used by the *Caulimoviridae*, which infect plants.

In the virion, the hepatitis B genome is double-stranded for 50–85% of its 3200-base length, with a full-length negative strand and a shorter postive strand of variable length. Entry to the target cell is mediated by the immunoglobulin A (IgA) receptor on the cell surface. Once within the target cell, the genome is converted to a fully double-stranded form by virion enzymes and the uncoated nucleocapsid enters the nucleus. Following this, a full-length positive sense mRNA is synthesized alongside multiple, overlapping mRNAs. At least three types of mRNA are produced. The mRNAs are translated to produce the four known hepatitis B proteins; C (core), P (polymerase/reverse transcriptase), S (surface) and X (*trans*-activator). The full-length RNA is copied into a full-length negative sense DNA by the viral P protein. Why the virus has such a complex replication cycle, involving the use of a viral reverse transcriptase (which uses over half of the coding capacity on the viral genome) rather than one of the cellular DNA-dependent DNA polymerases, is not clear. The genome is encapsidated within a core of C protein. The 1700–2800-base positive sense DNA strand is transcribed

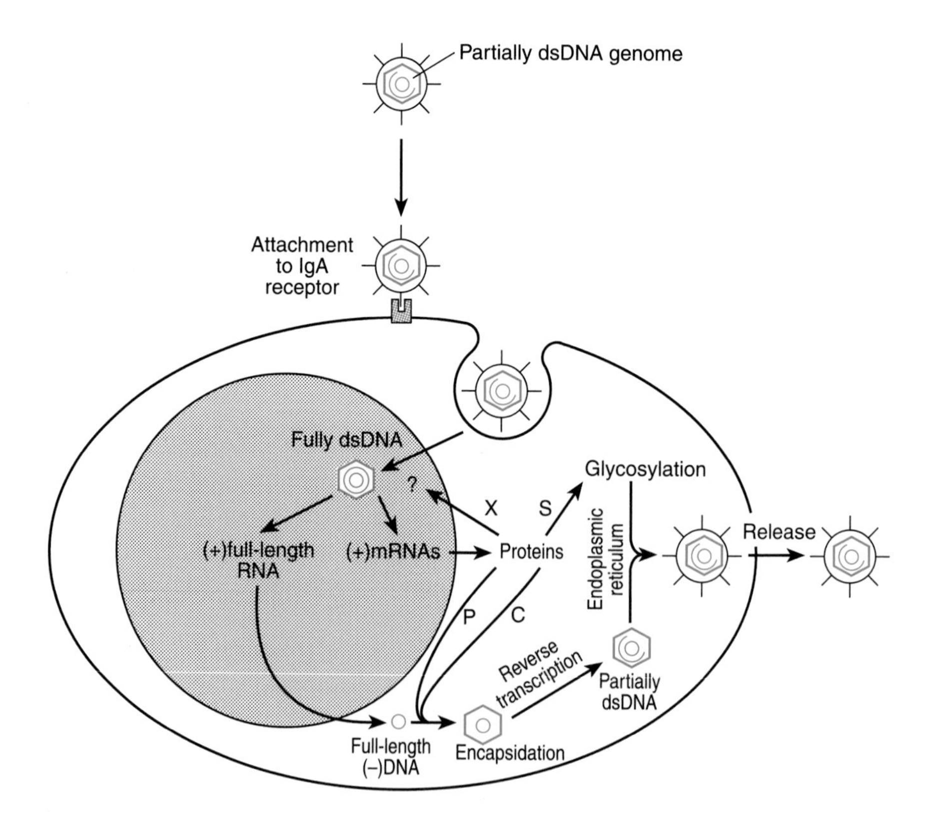

Figure 1.18: Hepatitis B virus replication.

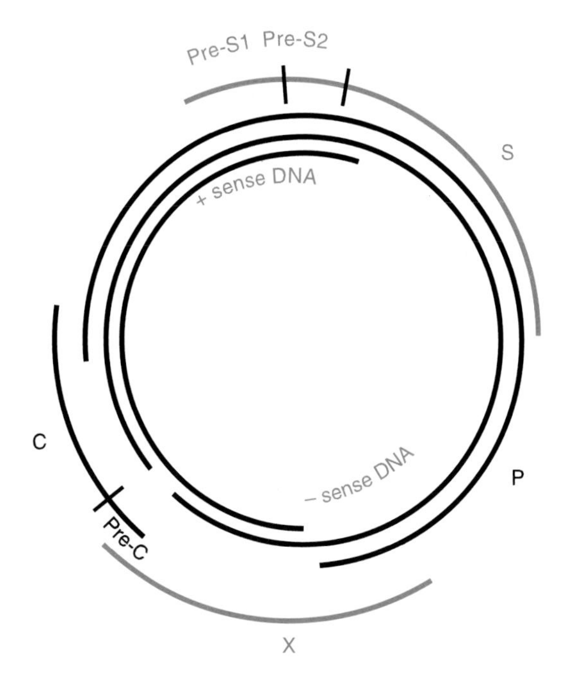

Figure 1.19: Organization of the hepatitis B genome, showing location of open reading frames.

from the full-length negative sense DNA only within the newly formed nucleocapsid, transcription ceasing before a full-length copy is produced, possibly due to exhaustion of the available nucleotides. The nucleocapsid is enveloped in lipid and S protein in the endoplasmic reticulum. The function of the X protein is not yet fully understood, although it appears to *trans*-activate some cellular genes and may be involved in the induction of hepatocellular carcinoma (see Section 1.7).

The *Hepadnaviridae* have very small genomes and make extensive use of overlapping reading frames, where a single region of DNA produces more than one protein. This means that a small amount of DNA can produce much more protein than it could by direct transcription and translation. In the case of hepatitis B, mRNAs are produced which start at the same point but vary in overall length. Of the four known genes (C, P, S and X), two (C and S) also produce multiple proteins using different start and termination sites. The positioning of genes on the hepatitis B genome is shown in *Figure 1.19*.

1.4.5 RNA virus replication

As with DNA, RNA is referred to as 'positive sense' (containing the mRNA sequences) or 'negative sense' (complementary to the mRNA sequences). Typically, viruses with RNA genomes have smaller genomes than DNA viruses, although there is some overlap with the smallest DNA viruses (especially the ssDNA viruses). The largest vertebrate virus RNA genome is about one-tenth the size of the largest DNA genome. One very important reason for this is that RNA replication is not usually 'proof-read' by the cell. DNA replication is very accurate since DNA polymerase enzymes also check the copied sequence and remove any mismatches, which are then replaced. This 'proof-reading' does not happen when RNA is produced, resulting in a vastly increased mutation rate (approximately 10^6-fold). For a complex organism which produces small numbers of offspring, mutation generally is to be avoided since most mutations are harmful. However, for a virus, the deleterious mutations will be less damaging since some of the vast numbers of progeny virus will be non-mutant viruses. The occurrence of beneficial mutations, while rare, will also be favored by the production of large numbers of progeny virus. Mutation is an important mechanism in helping the virus to evade immune surveillance (see Section 2.6). The mutation rate of RNA genomes is so high (one error in every 1000–10 000 bases copied) that it is unlikely that any copy of a viral RNA genome is exactly the same as the template from which it is copied.

The presence of a segmented genome in many RNA viruses also reflects the low fidelity of RNA replication, since one long molecule may be more likely to contain deleterious mutations. Segmented genomes also aid in the reassortment of RNA genes; while RNA can recombine (exchange sequences with a similar nucleic acid), most of the mechanisms to allow this are cellular, and are targeted at DNA. Segmentation provides an alternative route by which exchange of genetic information can occur. Influenza virus (*Orthomyxoviridae*) has seven (influenza C) or eight (influenzas A and B) segments of RNA making up its genome and is the best known and most studied of the viruses with a segmented genome. Other viruses with segmented genomes have from two to 12 segments. Influenza has a very high rate of genetic variation (antigenic drift), and also the ability to appear with new and very different genes (antigenic shift), possibly by exchanging RNA segments with influenza viruses from or in influenza-infected 'animal' reservoirs (*Figures 1.20* and *1.21*). Even RNA viruses with unsegmented genomes may vary rapidly. HIV is one of the most rapidly changing viruses, despite the fact that HIV has an unsegmented genome and also has two copies of the genome in each virion, which could moderate the effect of mutations. In fact, the use of reverse transcriptase seems to favor mutations of HIV, and the replication strategy used by HIV to allow full-length copies to be produced (copying a section of the genome from each genomic RNA molecule to produce a full-length copy) may also increase variation.

Replication of an RNA genome poses a particular problem for the virus. While many DNA viruses do produce polymerase enzymes, for such viruses these are one of the 'optional extras' in that the very small DNA

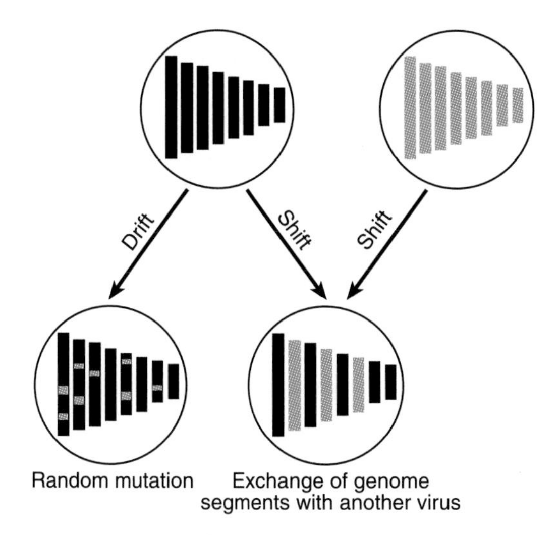

Figure 1.20: Principles of antigenic drift and shift. Colored bars represent novel RNA sequences.

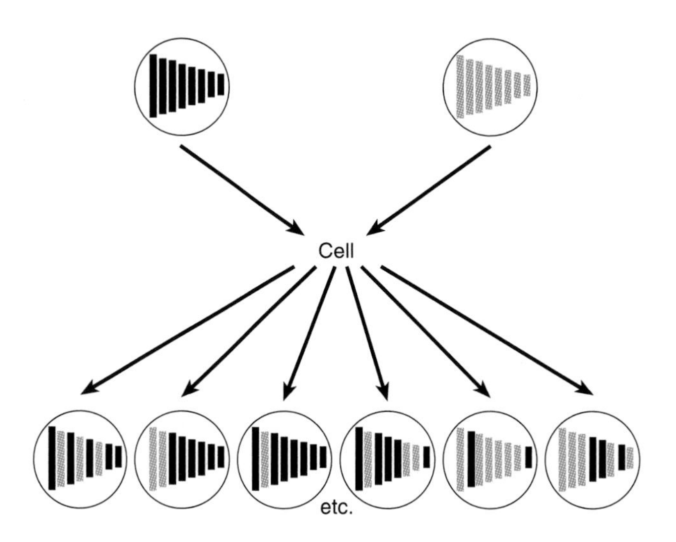

Figure 1.21: Principles of antigenic shift.

viruses can use cellular polymerases (when available). However, very few viruses with RNA genomes can use cellular enzymes, since cells do not contain useful levels of RNA-dependent RNA polymerase. This means that even very small RNA viruses need to produce a polymerase enzyme or find some way to use a cellular enzyme not originally intended for that purpose.

Viral RNA genomes may be double- or single-stranded. Where the genome is single-stranded, there may be positive sense RNA genomes, negative sense, or a mixture of both on the same molecule (ambisense).

Viruses with dsRNA genomes Reoviruses have been studied extensively and provide a model for dsRNA viruses in general (*Figure 1.22*). After entering the cell, reoviruses are digested by endosomes (they require the low pH digestion step to infect the cell) and then some are released by an unknown mechanism while others remain in the lysosomes. The digested core is the transcriptionally active form. There is some evidence that extracellular digestion can also activate the transcriptase activity. RNAs are made in the cytoplasm by a polymerase within the core. Only negative sense RNA is transcribed, producing positive sense mRNAs which are released from the core while the parental RNA stays inside. These mRNAs may be produced as two temporal groups, with synthesis of the second group requiring inactivation of a cellular control. The level of each viral protein is regulated both by transcriptional and translational controls, thus varying both mRNA level and mRNA activity. The released mRNAs are used both as mRNAs for protein synthesis and as virion components that are 'encapsidated' into virion precursors. How all 10 segments are correctly inserted is not known, but (unlike influenza) such sorting does appear to occur. Once within the new virus, a single round of negative sense RNA copies are made from the mRNAs, and these are then used to produce yet more mRNA. The genomic positive RNA strand actually *is* an mRNA, complete with the 5′ 'cap' modification (a 5′–5′ linked methylguanylate group, required for mRNA translation)

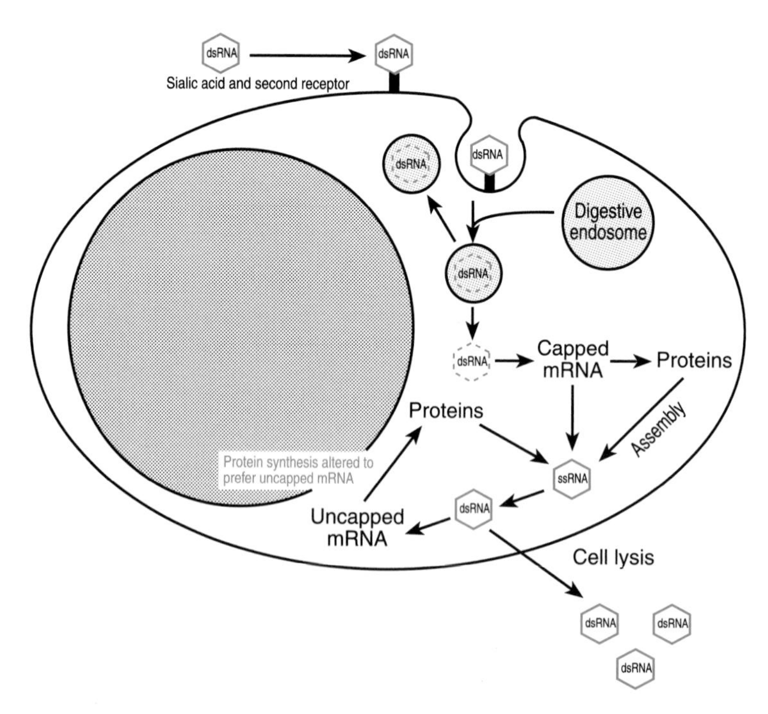

Figure 1.22: Reovirus replication.

seen exclusively on eukaryotic mRNAs, although it lacks the 3′ poly(A) tail seen on almost all eukaryotic mRNAs. While the cap of an mRNA is involved with forming the initiation complex for translation to protein, reoviruses alter cellular biochemistry so that uncapped mRNAs are preferentially translated late in infection, at the expense of capped mRNAs. Since all cellular mRNAs *are* capped, this 'turns off' cellular protein synthesis. However, the mRNAs produced from progeny reoviruses have no cap structure, and are translated preferentially. Reoviruses are very efficient at shutting off cellular synthesis, and specifically inhibit DNA, RNA and protein synthesis. Mature reovirus is released by cell lysis.

Viruses with positive strand ssRNA genomes. A positive sense RNA genome is infectious in itself, since it can code for the production of viral proteins by functioning as an mRNA. There are many viruses with this type of genome, the best-studied example being poliovirus (a member of the *Picornaviridae*) (*Figure 1.23*). Since the genome can act as an mRNA, there is no need for the virus particle to contain an RNA-dependent RNA polymerase. The viral genome can be translated to produce the polymerase and other necessary enzymes after entry to the cell.

In a superficially simple way, the poliovirus genomic RNA is translated to produce a single protein of more than 200 kDa. While it lacks a cap (see above), a specialized region of the RNA, the internal ribosomal entry site (IRES) binds directly to the ribosome. This allows

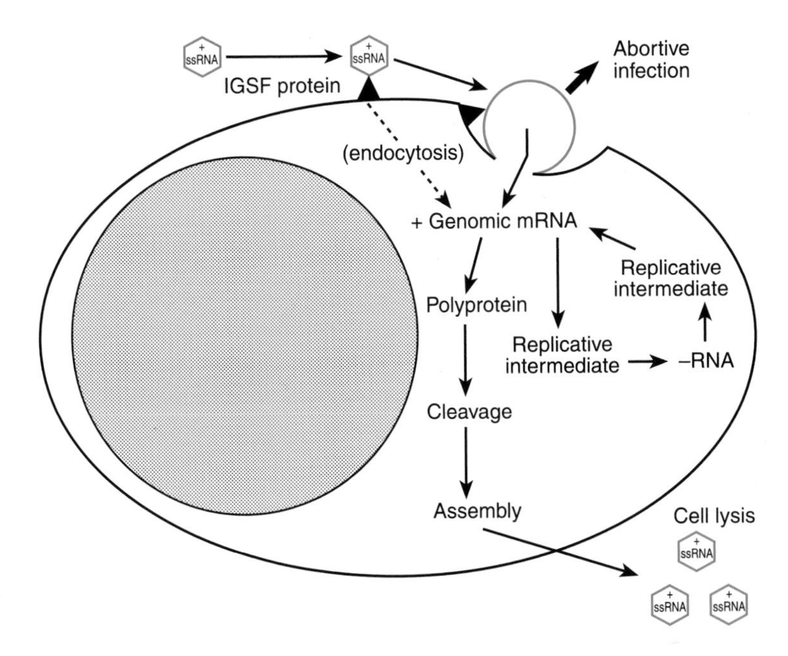

Figure 1.23: Poliovirus replication.

poliovirus to suppress cellular protein synthesis by inactivating a complex essential for translation of capped mRNAs.

Once synthesized, the 'polyprotein' is then repeatedly cleaved to produce not only the structural proteins that make up the virion, but also all of the enzymes necessary for replication, including the polymerase (*Figure 1.24*). However, one problem with such a simple scheme is that the virus produces equivalent amounts of every protein, even enzymes such as the polymerase which are not needed in the same amounts as proteins such as the structural subunits of the capsid. The polymerase then produces full-length negative sense RNAs, which in turn are copied into positive sense RNAs. Synthesis of RNA occurs in a partially dsRNA called a 'replicative intermediate' (RI) (*Figure 1.17*) associated with the smooth endoplasmic reticulum. Basically, this is one full-length molecule on which multiple copies of the opposite strand are being synthesized. The matching strands stay together until they are displaced by the next polymerase. Some of the positive sense RNAs are translated, while others are packaged as viral genomes by the newly synthesized coat proteins

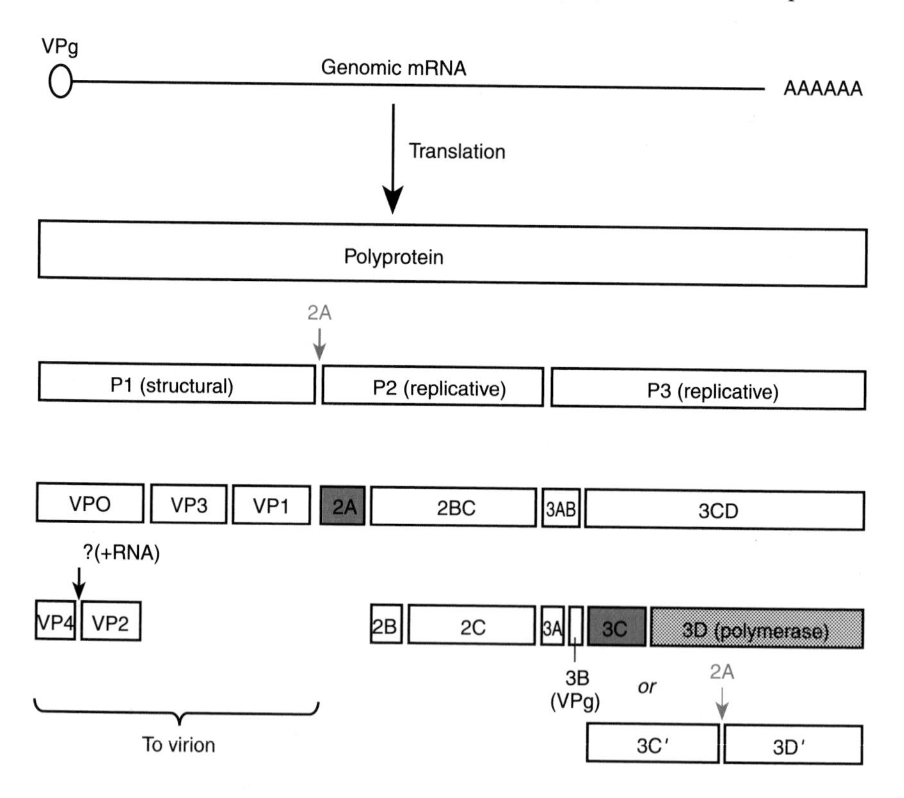

Figure 1.24: Poliovirus polyprotein cleavage. Proteinase 3C carries out all cleavages except those marked 2A or the latter, ? is carried out by an unknown proteinase after capsid assembly around the genomic RNA. Proteinases are shown in orange.

(which are still being cleaved even as assembly occurs). One copy of a small protein (VPg) is attached at the end of the genomic form of the positive RNA. VPg is important in viral RNA synthesis, where it appears to act as a primer. VPg is also associated with controlling the use of individual viral RNA molecules, with VPg attachment favoring a genomic role rather than function as an mRNA. Assembled poliovirus exits by lysing the cell.

Other viruses with positive sense ssRNA genomes may produce more than one mRNA, allowing greater control of the production of individual proteins, and some exhibit limited temporal control, with early (replication) and late (structural) proteins being produced at different times in the replicative cycle.

Viruses with negative strand ssRNA genomes. ssRNA genomes complementary to mRNA cannot function as mRNAs, and require a polymerase to produce mRNAs before any viral use of the cellular synthetic machinery can occur. The RNA-dependent RNA polymerase is carried in the nucleocapsid, and for the *Paramyxoviridae* this enzyme is the size of all the other proteins put together. Once within the cell, both mRNAs and genomic copies are made. During this process, a variety of dsRNA intermediates are seen (RFs and RIs) (*Figure 1.17*) as described previously, and a range of viral mRNAs are each translated to make different viral proteins, rather than a polyprotein.

Given the lack of complexity inherent in their small size, most RNA viruses do not have complex control mechanisms. However, all of the viral proteins are not produced to the same level, and a number of control mechanisms are used. For example, in replication of the *Paramyxoviridae* and the *Rhabdoviridae*, initiation of transcription occurs at a single promoter at the 3′ end of the genome, and is followed by progressive attenuation at the start of each new gene, so that genes farthest from the promoter produce the lowest levels of mRNAs. The polymerase is furthest away, while the (structural) nucleocapsid protein is closest and therefore made (as it is needed) in the largest amounts. When the capsid protein is present in sufficiently large amounts, it assembles around the forming RNA, preventing the attenuation and switching synthesis to full-length copies of the genome. Other virus proteins have important roles; notably inhibition of transcription by the M (matrix) protein, which is also responsible for mediating the binding of nucleocapsid to the modified membrane pre-envelope during assembly. Such multiple functions for an individual protein are typical of RNA viruses, reflecting the small size of their genomes.

In the case of influenza virus (*Orthomyxoviridae*) (*Figure 1.25*), the virus actually requires cellular RNA synthesis, since it cannot make the 5′ cap structure. The influenza virus has a special function which removes caps from nascent cellular mRNAs and incorporates them into viral mRNAs. Unusually, for a virus with an RNA genome, this requires the

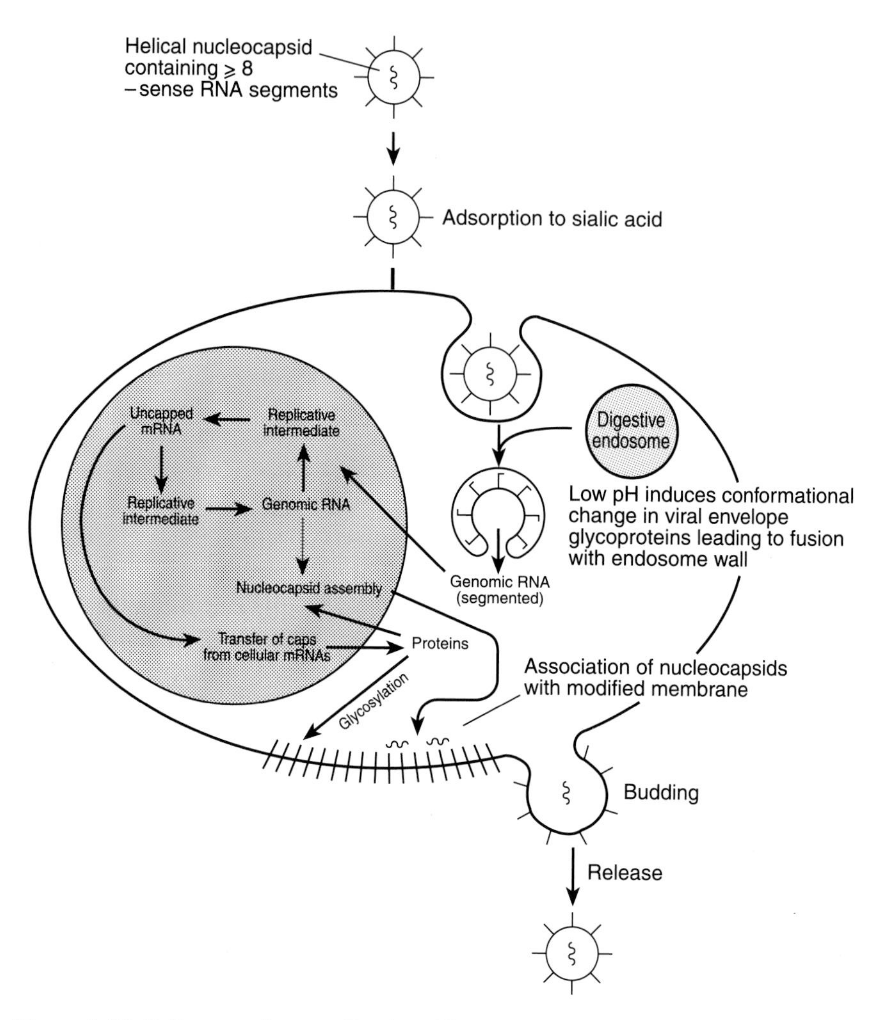

Figure 1.25: Influenza virus replication. Note: nuclear stage is not required for most negative sense RNA viruses.

viral mRNAs to be synthesized in the nucleus to allow capping, followed by transport to the cytoplasm for translation. This rather complex process appears to be unique to influenza virus, and may represent a way of down-regulating cellular protein synthesis. The influenza viral polymerase is also required to make full-length copies of the genome which are in turn used to make progeny genomes. Influenza shows another feature common to viruses with small genomes in that some of the gene segments code for multiple proteins using overlapping reading frames and splicing of mRNAs. Some of the *Paramyxoviridae* also use such methods, producing multiple mRNAs from the P gene.

Unlike the great majority of other ssRNA viruses of vertebrates, influenza also has the problem of its segmented genome. While this can assist with regulation of mRNA levels (and is important in influenza epidemiology), the virus has to ensure that all eight segments are included in the virion. No clear sorting signals are apparent, and it appears that, rather than sorting the genome segments, the virus may rely on encapsidating 12 or more RNAs at random, with only those containing the 'full set' being fully infectious. If only eight segments were encapsidated randomly, only 0.2% of virions would contain a full set. Encapsidation of 12 segments would give approximately 10% of virions with the full set, rising to approximately 15% with 14 segments encapsidated. However, many target cells may be infected by more than one virion, increasing the chances for successful infection.

Some groups of viruses have ambisense genomes (some regions positive sense, others negative sense). These include the *Arenaviridae* and the *Bunyaviridae* (both of which have segmented genomes), and viruses with ambisense genomes appear to replicate in a manner broadly similar to the negative strand viruses, using a viral polymerase to make mRNAs rather than having a genome that is directly translated.

ssRNA (positive sense) viruses that use a DNA intermediate. The *Retroviridae* are the only members of this class, and they have a very distinctive replication cycle, illustrated by that of HIV (*Figure 1.26*), although details differ between retroviruses. Uniquely, retroviruses synthesize RNA genomes from RNA genomes via a dsDNA intermediate. This 'reverse flow' of information from RNA to DNA gives the retroviruses their name ('*retro*' in Latin translates to 'backwards' in English) and also means that retroviruses must contain an enzyme able to carry out this reaction. This enzyme, reverse transcriptase, is central to the unique replicative cycle of the *Retroviridae*. Retroviruses are also the only diploid viruses, since each virion contains two copies of the viral genome. Full-length retroviral genomes may be copied from both copies of the genome, with the transcriptase jumping the gap between them, although some models suggest this may not be necessary.

On entering the cell, the retroviral genome is reverse-transcribed into a dsDNA intermediate. However, rather than existing as a circular DNA molecule free within the cell (an 'episome'), the retrovirus dsDNA appears to integrate into the cellular DNA using a viral integrase and remain there, being transcribed and replicated by the cell itself. In this state it is referred to as provirus. The proviral DNA can be expressed very efficiently and give rise to a productive infection, or it can remain 'silent'. This situation prevents clearance of the virus by the immune system since a silent integrated DNA presents no targets for the immune system. Even inactive integrated provirus can have significant effects on the cell, in particular by effects on nearby genes. It is by this mechanism, as well as by expression of viral oncogenes (see Section 1.7), that the induction of retrovirus-associated cancers can occur. However, the retroviral genome is small, and oncogenes in the retroviral genome are present in place of

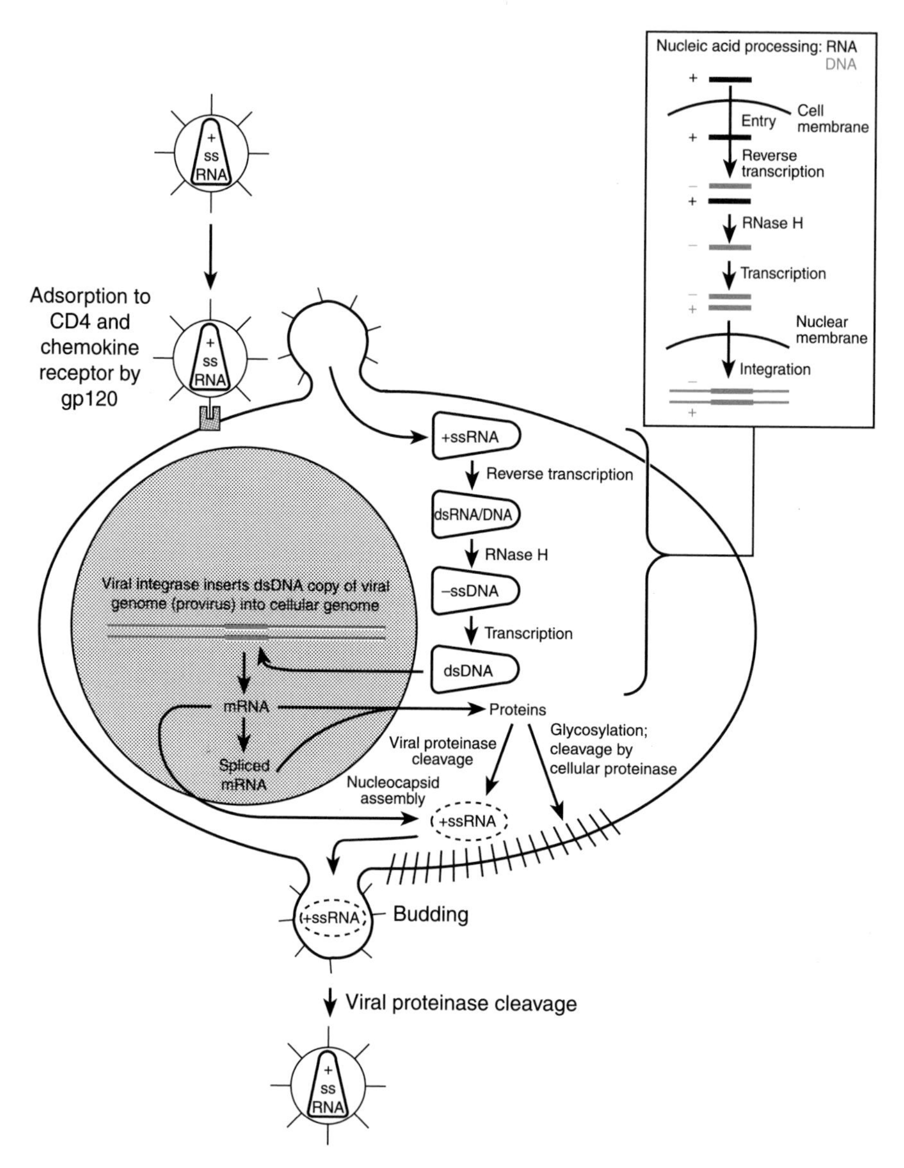

Figure 1.26: HIV replication. In latent infection, mRNAs for structural proteins are not produced.

essential genes in almost all retroviruses. Thus, a retrovirus carrying an oncogene requires a co-infecting complete 'helper' virus (see Section 1.8.4). This requirement casts serious doubt on the ability of such viruses to cause cancers outside the laboratory, and it is likely that integration-based effects are more relevant in 'wild' retroviruses. As with all other viruses, once the retroviral genome has made copies of itself and has made the viral mRNAs which in turn have made the viral proteins, these components are assembled into the virus particles, which are

Table 1.3: Proteins of the human immunodeficiency virus type 1

gag	Polyprotein, cleaved by the viral aspartyl proteinase, produces the core proteins
pol	Polymerase/reverse transcriptase, RNase H, aspartyl proteinase and integrase
env	Polyprotein (gp160), cleaved by cellular enzymes to produce gp41 and gp120
Auxiliary genes	
vif	Required for production of infectious virus, modulates viral transport
vpr	Regulates gene expression, regulates cell cycle
tat	Up-regulates viral mRNA synthesis by binding to the TAR (*trans*-activator response element) in the transcripts from the long terminal repeat
rev	Required for mRNA transport and processing
vpu	Virion release, CD4 degradation
nef	Controls viral gene expression, binds and down-regulates CD4, down-regulator of viral metabolism in some cell types

released from the cell. In the case of retroviruses, this is by budding from the cell surface.

The retroviral genome contains a long unique sequence between two long terminal repeats (LTRs) and contains three main genes, *gag* (core proteins), *pol* (polymerase/reverse transcriptase) and *env* (envelope glycoproteins). In the human retroviruses, many small regulatory or 'auxiliary' proteins are also produced (*Table 1.3*). Analogs of these proteins may be present, but are not as well characterized in animal retroviruses.

One very notable feature of retrovirus replication is the extremely rapid mutation rate of the virus. This enables it to evade the immune system by changing the targets for the immune response. However, since the virus is diploid, mutations should be moderated. Some hypotheses exist to explain the high rate of retrovirus mutation.

(1) Reverse transcriptase is a very low fidelity polymerase. Combined with the high rate of mutation inherent in RNA genomes, this would ensure that almost all copies of the HIV genome were different in some way from their parent molecule.
(2) 'Jumping' between genome copies, if it occurs, could introduce errors, and could also mean that the transcriptase is prone to jumping elsewhere, particularly at specific sequences.
(3) Where multiple retroviruses infect a cell, the genomes might be able to reassort, giving one copy of the genome of each 'parent'. It is thought that the close interaction between the two genomes within the virion requires that they be very similar, so while this might occur with viruses of the same 'species', it would probably not occur between widely different viruses.
(4) Since retroviruses produce the specific enzymes required for integration and excision of their DNA, these may give a far higher rate of recombination than normally occurs. Since cells appear to contain a significant number of endogenous retroviruses or (see Sections 1.7 and 1.8), it is possible that recombination with these could occur, increasing the resultant variation.

Hypotheses 3 and 4 are speculative, but could help to account for the mutation rate of retroviruses. While several human retroviruses have now been identified, it is clear that many retrovirus-related elements are present in the genomes of most higher organisms, and the role of these elements is not yet known. Some appear to be only fragments, while others can move around inside the cell. There is a lot still to be learned about the nature and effects of these elements.

1.4.6 Virus assembly

In general, more complex virus structures require more complex assembly pathways. Poliovirus provides an example of a 'simple' virus. The polio virion contains five proteins, four of which make up the capsid (see Section 1.3.3). If produced in a cell-free translation system, these four proteins can assemble themselves into a capsid. In the infected cell, the proteins assemble themselves into precursor forms (pentamers) which are then used in capsid formation. The genomic RNA is either inserted into preformed capsid or condenses with the proteins during capsid formation, although the available evidence favors the latter. Adenoviruses have similar (icosahedral) capsid morphology to poliovirus, but have much larger genomes and a far more complex structure (see Section 1.3.3) and pathway of assembly. Additional 'scaffolding' proteins (proteins required for assembly but not present in the final form) are required at two stages. A large (100 kDa) protein is required for the formation of the hexamer subunit (itself a trimer), while two further scaffolding proteins are required to actually assemble the capsid, into which the DNA is inserted.

Once assembled, the viruses must leave the cell. In the case of many bacteriophages, this simply involves the cell bursting as a result of the virus infection. In eukaryotic cells the situation is usually more complex. Viral protein synthesis can only take place in the cytoplasm, and the majority of viruses are assembled there. With some viruses, such as the *Herpesviridae* or the *Orthomyxoviridae*, viral proteins migrate to the nucleus, and are there assembled into nucleocapsids. Enveloped viruses then acquire their membrane component from a specific cellular membrane. The plasma membranes is a common source, but in fact a variety of cellular membrane may be used, depending upon the nature of the infecting virus. In the case of *Paramyxoviridae*, the chemical composition of an area of the plasma membrane is altered so that it becomes more rigid, and viral glycoproteins are inserted into this area of membrane. A specific viral sub-membrane protein accumulates on the inside of the membrane at this point and mediates interaction with the completed nucleocapsid. The whole complex then pouches out of the plasma membrane and buds off into the extracellular environment.

Retroviruses are another group of viruses that exit from cells by budding. The nucleocapsid is assembled either in the cytoplasm or at the plasma membrane (depending on the specific retrovirus), and the core then buds out of the cell surrounded by a section of modified plasma membrane. Budding of a retrovirus is shown in *Figure 1.27*. Other viruses acquire membranes internally, from any of a variety of sources, and then move to the cell surface, often in cellular vacuoles which fuse with the plasma membrane to release the virus (*Figure 1.28*). Non-enveloped viruses are usually released by cell lysis although, since some produce infectious virus without lysing the cell, a transport mechanism, probably involving vacuoles derived from cellular membranes, must exist. 'Bursting' (i.e. non-active lysis) of a dying cell does occur and can also produce infectious virus particles.

1.5 Effects of virus infection on the host cell

Many viruses produce specific effects on the host cell, mediated by viral functions. An excellent example is the herpes simplex virus vhs (virion host shut-off) protein, carried in the tegument of the virion. The vhs function destabilizes cellular mRNAs, causing them to be degraded. Unsurprisingly (given their complex nature), the herpesviruses are also thought to use other mechanisms of inhibiting cellular synthesis, including alterations to the phosphorylation state of cellular enzymes.

Many other viruses also decrease cellular synthesis. This can involve competition for the cellular protein synthesis machinery by the production of massive amounts of viral mRNA, as seen with vesicular stomatitis virus (VSV), the best studied of the *Rhabdoviridae*. Some viruses alter cellular protein synthesis so as to favor the translation of viral mRNAs. This is seen with reovirus where the uncapped viral mRNAs produced late in infection are translated preferentially. Cellular RNA and DNA synthesis are also inhibited by reoviruses. In some cases, there can be almost total shut-off of cellular synthesis, but this is rapidly toxic to the cell, and some viruses use a more limited inhibition allowing a long and productive infection.

It is important to remember that an ideal parasite does not kill its host, so it may be favorable to the virus to moderate shut-off of cellular functions. Culturing cells chosen for susceptibility to virus infection in the absence of any immune protection is a very unnatural system, and the acute, lytic infections often seen in cultured cells may be a poor reflection of events occurring during viral infection of the whole organism.

1.6 Types of virus infection

In cultured cells, some viruses are only capable of an active, 'acute' infection, where the host cell is rapidly killed as a result. However, many others can infect a cell and actively produce virus without immediate cell death by using less destructive methods of exit, such as budding from the cell surface. Alternatively, the virus may become latent within the cell, as

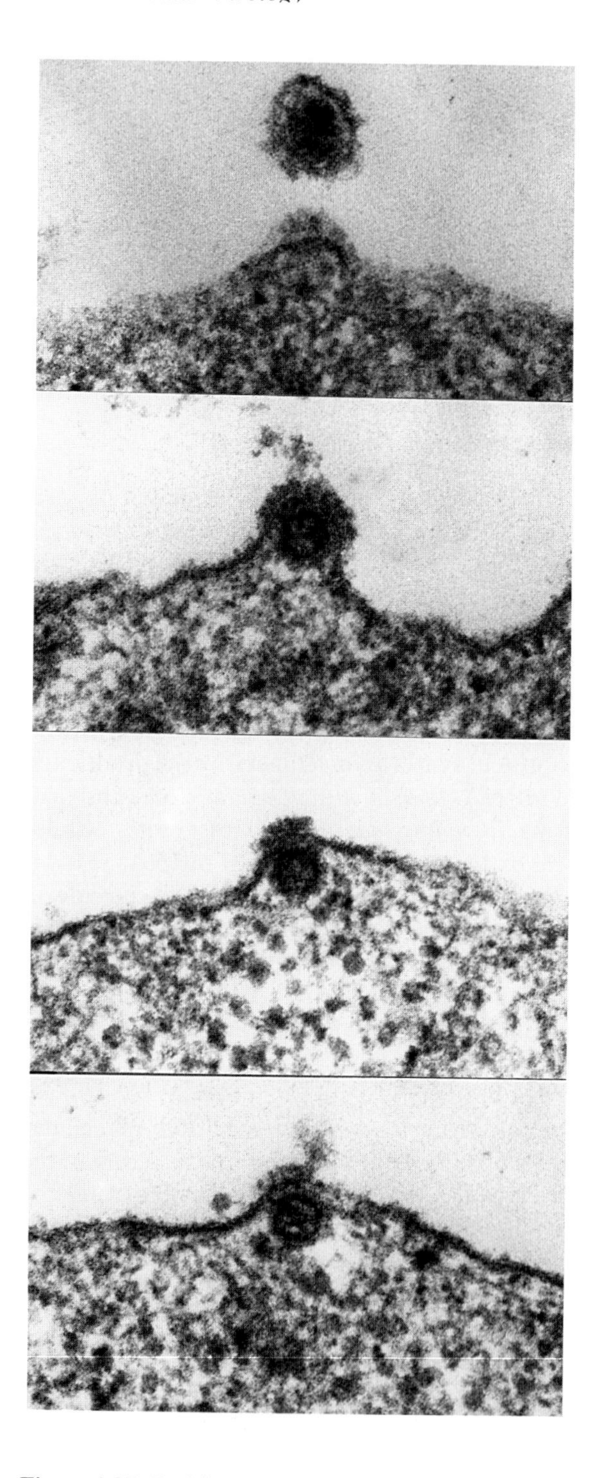

Figure 1.27: Budding of a virus from the cell (baboon retrovirus). Photographs courtesy of Dr Ian Chrystie, Department of Virology, St Thomas' Hospital, London.

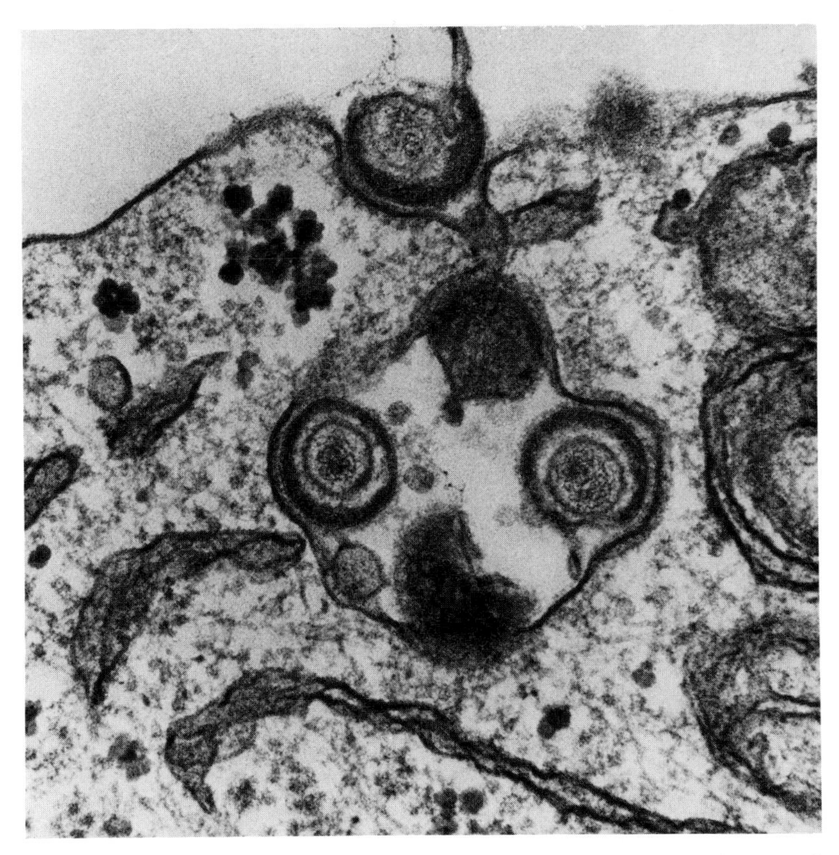

Figure 1.28: Vacuolar transport of a virus (varicella-zoster virus). Photograph courtesy of Professor Charles Grose, Department of Pediatrics, University of Iowa Hospitals, Iowa City, Iowa, USA.

mentioned earlier, either integrating with the chromosome or remaining as a self-replicating extrachromosomal nucleic acid (an episome).

Viral infections at the cellular level can be subdivided into: acute/lytic; persistent/chronic/slow; latent/proviral; or transforming/oncogenic. The effects of such infections are shown in *Figure 1.29*.

An acute or lytic infection results in rapid cell death, and is commonly seen in virus infection of cultured cells. Despite this, it appears that such rapidly cytotoxic infections may be less common *in vivo*, where interactions with the immune system moderate infection. Lysis of infected cells may be an active process involving viral enzymes, or simply a reflection of indirect damage to the cell resulting from virus infection. Lysis of the cell has the effect of releasing virus, which can then initiate further infections.

A low level infection which kills only a small proportion of cells has been observed in cell culture with several types of virus, although an external limitation on infection, such as antiviral drugs or cytokines, is often necessary to prevent a lytic infection. Examples include the use of interferon or antiviral drugs such as aciclovir (for herpesviruses).

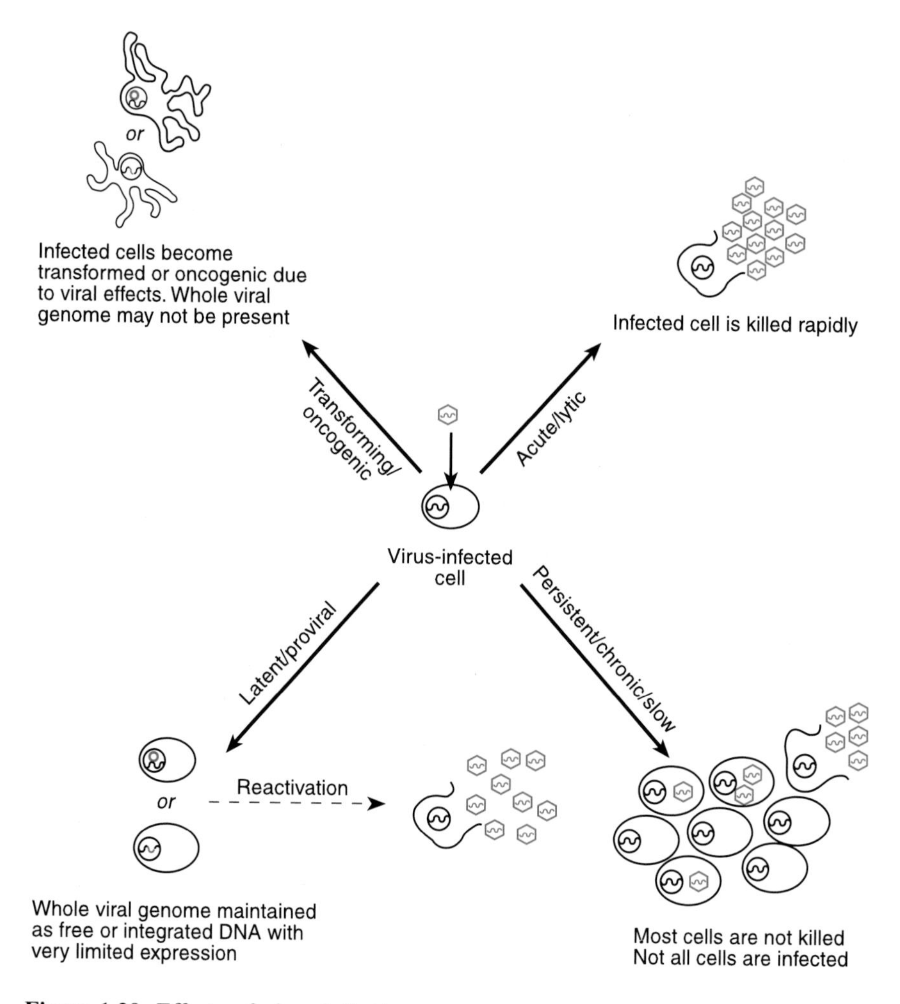

Figure 1.29: Effects of virus infection.

Infections of this nature appear to be relatively common *in vivo* following infection with a range of viruses, since many elements not present in cell culture (notably the effects of the immune system) act to limit virus replication. The terms chronic and persistent are not distinct. In such infections, virus replication is at low levels, and the immune response may be moderated by the development of 'tolerance' to viral antigens, allowing a long-term, low-key infection to become established. The type of cell infected may also play a major role, since some cell types will not permit full virus replication. A commonly quoted example of a persistent infection in humans is the rare sub-acute sclerosing pan-encephalitis (SSPE) which can follow infection with measles virus. In SSPE, virus replicates at low levels without producing infectious virus with altered production of viral proteins and an atypical immune response. This and other persistent infections may also be moderated by 'defective interfering' viruses (see Section 1.8.4). Probably due to measles virus immunization programs, SSPE is becoming increasingly rare, but it does represent a good example of a virus continuing to replicate at a very low level over a long period of time. An alternative example of prolonged virus infection is that of 'slow' viruses, where the virus is present and replicating at low levels for a long time before producing apparent disease. Examples in humans include JC virus (a polyomavirus) and the transmissible spongiform encephalopathies (see Section 1.8.6).

In a true latent infection, the virus is present but is almost totally inactive. Cellular factors as well as viral factors are believed to be important in controlling latency, acting to block functional expression of virus regulatory proteins. The defining characteristics of a latent infection are that only part of the genome is active, and no virus replication occurs. The most quoted example of such an infection is that of herpes simplex virus (HSV) where, following the acute infection, the virus becomes latent in ganglionic nervous tissue. Under these conditions, only a very small region of the viral genome is active, producing three RNA 'latency-associated transcripts' which appear to be involved in reactivation from the latent state. The viral genome appears to be maintained as an episome. There is no firm evidence of the production of a viral protein by latent HSV, but this must be compared with the production of about 70 proteins (and corresponding mRNAs) during an acute infection with this virus. Clearly, while latent, the virus is effectively 'silent', presenting no targets for the immune system and producing very limited (if any) effects on the host organism.

An alternative latency is that of proviral integration, seen with retroviruses, where the genome is copied into DNA and integrated into the cellular chromosomal DNA. This has parallels in the 'lysogeny' of certain bacteriophages (such as bacteriophage lambda) which can integrate their DNA into the bacterial chromosome. Integrated DNA is replicated along with the host cell DNA, and may (as noted above) account for a significant percentage of cellular DNA. It is estimated that 10–15% of murine DNA may be integrated retroviral genomes replicated along with the cellular DNA, and multiple retrovirus-related elements have been identified in humans. The role of endogenous retroviruses (if any) is uncertain, but there have been numerous suggestions including involvement in the pathogenesis of transmissible spongiform encephalopathies (see Section 1.8.6) and the production of localized immunosuppression to prevent rejection of 'foreign' placental tissue during pregnancy in mammals.

1.7 Viral transformation and oncogenesis

An alternative result arising from the integration of retroviral DNA is transformation of the infected cell, the effects of which are shown in *Table 1.4*. While it appears that transformation is linked to oncogenesis, the production of cancers *in vivo*, many other factors are involved. These include the correct functioning of immunological monitoring systems

Table 1.4: The nature of transformation

Properties of transformed cells
- 'Immortalization' of cells from some species (including human and mouse, but not chicken cells)
- Loss of regulation of cell growth (cell density inhibition, growth factor requirements, requirement for solid substrate)
- Changes to cell appearance and structure (cytoskeleton, expression of unusual surface proteins, reduced adhesion, proteinase secretion)
- Abnormal chromosome numbers (aneuploidy)
- Altered patterns of transcription
- Where cells are transformed by viruses, the viral genome is present, usually integrated into the cell genome. Specific viral oncogenes may be expressed and/or relevant cellular functions altered.
- Production of transforming growth factors (TGFs)

Analogies with cancer
- Cancer cells show unregulated growth without normal limits
- Cancer cells express unusual surface proteins
- Cancer cells may show chromosomal abnormalities, including the insertion and expression of viral nucleic acid
- Some transformed cells can induce cancers in animals

BUT
- Not all transformed cells induce cancers in animals – other elements (particularly the immune system) appear important
- Not all viruses that can transform cells are oncogenic

CONCLUSION
- Transformation is not by itself enough to cause cancer, but appears to be an early step in a multi-stage process

which normally ensure that transformed cells are destroyed, notably the cytotoxic T-cell response. Also very important at this stage are the activities of the 'tumor suppressor' genes of the host cell such as the *p53* or retinoblastoma (*Rb*) genes. For example, the *p53* gene appears to be involved in apoptosis (programed cell death; see Section 2.4); a cell which has lost *p53* function will not obey the normal controlling mechanisms (basically, commands to kill itself) and can grow out of control. It also appears that *p53* regulates the cell cycle via p21 protein (also known as CIP1 or WAF1), an inhibitor of cyclin-dependent kinases, and that *p53* is involved in DNA repair. Activity of *p53* may be regulated by cellular suppressor proteins and by phosphorylation of the p53 protein by cellular kinases. The Rb protein is important in regulating the cell cycle, and also interacts with the cyclin proteins. As with *p53*, mutation of the *Rb* gene is linked to the development of cancer.

With retroviruses, transformation can result from the presence within the integrated DNA of an oncogene (cancer-associated gene), frequently a cellular gene carried in a defective viral genome. Alternatively, integration of the DNA can alter the regulation of a nearby cellular oncogene, producing similar effects. It is important to realize that such transformation can occur even when the virus DNA is inactive, and that the initial difference between stable integration and oncogenic integration may be

Table 1.5: Factors which may be involved in virus-induced transformation

Viral nucleic acid integrating into cellular tumor suppressor genes, inducing transformation
Expression of viral oncogenes (of various types)
Expression of viral oncogenes as partial or fusion proteins, freed from normal controls
Viral controls altering expression of cellular genes, either nearby (*cis*-acting) or by *trans*-acting factors
Chromosomal translocations (altering expression of cellular genes)

very small. Other viruses may also be transforming and/or oncogenic by a range of effects (*Tables 1.5–1.7*). Herpesviruses, papovaviruses and hepadnaviruses are all associated with transformation of cells and with oncogenesis, while adenoviruses are transforming but not known to be oncogenic in humans. With many transforming viruses such as the adenoviruses and papovaviruses (both polyomaviruses and papillomaviruses), viral regulatory (early) proteins may affect cell growth directly. These and other effects are summarized in *Table 1.7*.

It is now clear that viruses are very important in a range of human and animal cancers, but that, in many cases, other factors (known and unknown) are required for oncogenesis. However, prevention of viral infection by vaccination provides a real hope for preventing some cancers, notably those

Table 1.6: Viruses implicated in cancer

Virus family	Human	Animal	Tumor types	Associated human cancers (agent)
Adenoviridae	No	Yes	Solid tumors	None known
Hepadnaviridae	Yes	Yes	Hepatocellular carcinoma	Hepatocellular carcinoma (hepatitis B virus)
Herpesviridae	Yes	Yes	Lymphomas, carcinomas and sarcomas	Nasopharyngeal carcinoma (EBV) Burkitt's lymphoma (EBV) Other malignancies including gastric cancer, some Hodgkin's lymphomas and B-cell lymphomas (EBV) Kaposi's sarcoma (HHV8) Body cavity lymphoma (HHV8) Multiple myeloma (HHV8)
Papovaviridae				
Papillomavirus	Yes	Yes	Papillomas and carcinomas	Papillomas (many HPV subtypes) Cervical carcinoma (limited range of HPVs, notably types 16 and 18)
Polyomavirus	(No)	Yes	Solid tumors	None known
Poxviridae	No	Yes	Myxomas and fibromas	None known
Retroviridae	Yes	Yes	Hematopoietic cancers, sarcomas carcinomas, leukemias	Adult T-cell leukemia (HTLV-1)

EBV, Epstein–Barr virus; HPV, human papillomavirus; HTLV, human T-cell leukemia virus.

Table 1.7: Mechanisms of transformation/oncogenesis in viruses infecting humans

Virus	Mechanism
Adenovirus (oncogenicity in humans not demonstrated)	E1A protein: interaction with Rb protein, interference with MHC-I antigen presentation E1B protein: interaction with p53 protein Fragments of viral genome integrate into host chromosomes
Epstein–Barr virus (herpesvirus)	Chromosomal translocations activating cellular oncogenes (notably the c-*myc* gene) EBNA5 protein complexes Rb and p53 Complete viral genome maintained as episome Co-factors: nitrosamines in diet (nasopharyngeal carcinoma); malaria and other factors (Burkitt's lymphoma)
Human herpesvirus 8 (herpesvirus)	K1 transforming protein G protein-coupled receptor K cyclin
Hepatitis B virus (hepadnavirus)	pX protein may *trans*-activate unknown cellular genes Insertional effects on cellular oncogenes including *p53*, cyclin A, retinoic acid receptor Forced hepatocyte replication due to virus-induced liver damage Oncogenesis linked with *p53* mutation and chromosomal changes All or part of viral genome may integrate into host chromosomes Co-factors: aflatoxins or nitrosamines in diet, hepatitis C infection
Hepatitis C virus (flavivirus)	Unknown (no molecular data available)
HPV 16, HPV 18, other HPVs (papillomavirus)	E6 protein: interaction with p53 protein E7 protein: interaction with Rb protein Other early proteins (effects on cell cycle) Insertional effects on cellular oncogenes Oncogenesis strongly linked with *p53* mutation Fragments of viral genome are inserted into host chromosome (non-oncogenic HPV genomes are maintained as episomes) Co-factors: smoking, possible link with HSV 2 infection
Polyomaviruses (oncogenicity in humans not demonstrated)	SV40 large T antigen protein: interaction with both p53 and Rb Fragments of viral genome insert into host chromosome
Retroviruses	Viral oncogenes (related to host cell oncogenes, carried in place of essential viral genes in defective viruses[a]) Insertional effects on cellular oncogenes HTLV-1 *tax* gene may *trans*-activate unknown cellular genes Complete viral genome inserts into host cell chromosome as essential part of viral life cycle

[a]Except Rous sarcoma virus of chickens, which carries an oncogene in a competent virus.

associated with the hepatitis B virus (for which a vaccine is available) or with Epstein–Barr virus (EBV, for which a vaccine is under development).

1.8 Sub-viral infections

With increasingly precise knowledge of exactly what constitutes a virus at the molecular level, it has become clear that there are various infectious agents which do not show the full spectrum of characteristics associated with most viruses. These range from replication-deficient 'satellite' viruses which need to use one or more elements provided by a replication-competent 'helper' virus, down to naked RNA (viroids) and (possibly) 'infectious' proteins (prions). There are also a range of defective viruses as well as mobile genetic elements seen in every organism from bacteria to humans which bear at least some resemblance to viruses.

1.8.1 Satellite viruses

A satellite virus can resemble a replication-competent virus in many ways and produces virus-specific structural proteins, but has adapted to require co-infection of the cell with a different but specific 'helper' virus, which allows the satellite virus to complete its replicative cycle. The only known satellite virus of humans is adeno-associated virus (AAV), a member of the *Parvoviridae*, which is closely related to the replication-competent parvoviruses. Despite this, it is unable to replicate without the presence in the same cell of a helper adenovirus (or, under some cicumstances, a herpesvirus). The helper function required appears to be a modification of the cellular environment, since some toxins can substitute for the helper virus function, but has not been characterized. If no helper virus is available, AAV is capable of integrating into the cellular genome until helper functions become available. AAV does not appear to be associated with any known human disease, but can interfere with some functions of the helper virus.

1.8.2 Virusoids

These are the 'next step down' from satellite viruses, and are associated with viruses of plants. No human examples are known. They have very small genomes (200–400 bases of circular ssRNA), and do not code for their own structural proteins. Rather, they 'steal' these from the helper virus. Virusoids replicate in the cytoplasm using helper virus functions.

1.8.3 Viroids

Viroids infect plants and produce a range of diseases. They are very small (240–375 bases) 'single-stranded' circular RNA molecules, with very pronounced secondary structure. In fact, even though there is only one strand of RNA, the vast majority of the viroid RNA is paired with other regions, forming a tightly coiled structure (*Figure 1.30*). This convoluted structure helps to stabilize them, since they must survive without a protective coat of protein in a world full of nuclease enzymes that would normally destroy a ssRNA very rapidly indeed. They are the smallest self-replicating infectious agents to have been characterized, but despite this (and unlike satellite viruses or virusoids) they can replicate without any requirement for a helper virus.

Figure 1.30: Base-paired structure of a viroid RNA. Reproduced from Sänger (1984) with permission from Cambridge University Press.

Viroids do not actually code for any proteins. The viroid genome is replicated in the cellular nucleus entirely by cellular enzymes, and there is no DNA or protein stage in their life cycle. The enzymes transcribing the viroid RNA are actually cellular DNA-dependent polymerases (actually RNA polymerase II), but the unusual secondary structure of the viroid RNA means that they are used as templates at high efficiency by this enzyme which normally makes RNA from DNA. Since the viroid RNA is produced as concatamers (polymers of the genome) and no viroid proteins are produced, how are the genomes cut to unit length? Surprisingly, this is not by a cellular protein, but rather is mediated by the viroid RNA itself, functioning as a *ribozyme*, a word derived by combining ribonucleic acid with enzyme. This ability of RNA to catalyze a reaction without the involvement of a protein was very surprising when it was first discovered in 1981, and has been studied intensively using viroids as a model system, although similar cleavage events occur in virusoids, satellite viruses and other systems. The mechanisms of ribozyme cleavage and their potential uses are discussed in Section 4.6.

Despite the fact that all known viroids infect only plants, they have certain similarities to the hepatitis delta virus (the delta agent or HDV). HDV is associated with more severe forms of hepatitis B infection, and has some characteristics of viroids, virusoids and satellite viruses. HDV is dependent upon a helper virus (in this case, hepatitis B) to form HDV particles (although the HDV genome can replicate independently), and uses hepatitis B structural proteins to form the HDV particle, along with the delta antigen produced from the single open reading frame on the HDV genome. The genome of HDV is larger than a viroid, a 1680-base circular, ssRNA. However, it has a viroid-like secondary structure in which 70% of bases are paired, and contains base sequences related to those seen in viroids but distinct from those seen in virusoids or satellite viruses. In addition, its replication appears to involve a viroid (or virusoid)-like self-cleavage of the RNA. From the case of HDV, it is clear that the arbitrary classification of satellite/virusoid/viroid actually covers a spectrum of related sub-viral agents.

1.8.4 Defective viruses

In any discussion of sub-viral infectious agents, the various replication-defective viral forms generated by replicating viruses must be considered. In the case of infuenza, as discussed in Section 1.4.5, virions which do not contain the 'full set' of genome segments could replicate if co-infecting viruses provided the missing functions, while retroviruses which have replaced part of their own genome with cellular oncogenes require a co-infecting competent 'helper' virus (see Section 1.4.5). Many viruses, particularly those with RNA genomes, produce 'defective interfering' (DI) particles, particularly at high multiplicities of infection. These contain only part of the viral nucleic acid which, being smaller, can replicate more rapidly than whole viruses, provided that a replication-competent virus is present in the same cell to supply the missing functions. DI particles are important in establishing persistent virus infections, since they can alter the course of infection by interfering with viral replication. All of these viruses are very similar to their 'parent', and (necessarily) exist as mixed populations. In the case of DI viruses and of oncogenic retroviruses where an oncogene replaces essential viral functions, they can have profound effects on the course of infection.

1.8.5 Mobile genetic elements and endogenous viral sequences

These were first studied in bacteria, where it became clear that some self-replicating genetic elements, or plasmids, could transfer from cell to cell. Some complex plasmids can actually initiate this transfer, and plasmid transfer, taking with it the genes coding for resistance to antibiotics, has caused significant clinical problems. Other mobile genetic elements (transposons) integrate into and can move within the genome of the bacterium.

Eukaryotic cells have direct equivalents of plasmids, known as episomes, similar to those seen in some virus infections. As with prokaryotes, there are other mobile genetic elements. In fact, the VLP–Ty system discussed in Section 3.3 is based on the Ty retrotransposon, a mobile genetic element related to retroviruses that uses a reverse transcriptase and a retro-transposon-coded protein to relocate itself within the genome of its host yeast cell. Such mobile elements fulfil at least some of the requirements of a sub-viral infection, and further extend the range of known sub-viral agents.

In addition, both prokaryotes and eukaryotes contain virus-derived sequences of varying degrees of completeness, many derived from lysogenic bacteriophages and integrated retroviral sequences respectively. In mammals, these elements may account for up to 10% of the genome. Possible functions for retroviral elements are discussed in Sections 1.7 and 1.8.6, and such elements have the potential to interact with both cellular elements and intracellular viruses. Clearly the spectrum of sub-viral agents extends all the way down to a few bases of virus-derived genetic material integrated into the cellular genome which may have some (even if very limited) potential to produce effects on the cell.

1.8.6 Prions

These agents appear to be quite different from the spectrum of sub-viral agents described above. Work during the 1960s showed that the agents causing spongiform encephalopathies in humans and animals were transmissible. Human diseases of this type are: kuru, a disease limited to Papua New Guinea and thought to be caused by preparation and ritual consumption of infected human brain tissue; Creutzfeld–Jakob disease (CJD), a rapid onset dementia observed worldwide with some genetically related clusters; Gerstmann–Sträussler syndrome (GSS), a dementia of longer duration, nearly always restricted to specific family groups; and fatal familial insomnia, another inherited disease. The best studied spongiform encephalopathy is scrapie, a disease of sheep. All are long incubation diseases causing severe localized neurological damage, producing holes in the brain which give it a sponge-like appearance.

Scrapie may be transmitted experimentally to other animals, and has been observed to transmit to captive animals fed with sheep-derived materials, notably deer and mink.

Despite much work in the 1950s and 1960s to identify a 'slow virus' causing spongiform encephalopathies, none were identified. In fact, the agent had many properties suggesting that it was a protein with no accompanying nucleic acid (see *Table 1.8*), most notably its resistance to inactivation. Infectivity appeared to be linked to a modified form of the cellular protein PrP^{C}. The modified form is known as PrP^{Sc}, and was present in fibrils within the brain associated with areas of neuronal damage. The term 'prion' was selected to describe such an agent, although this was not the first use of the word. In virology, a prion is now defined as a 'small proteinaceous infectious particle which resists inactivation by agents which destroy nucleic acid and contains an essential modified isoform of a cellular protein'.

The prion hypothesis was very much against the prevailing orthodoxy, where the use of nucleic acids to encode and copy the genetic information was not only an essential property of living organisms, but actually formed the definition of life, and an alternative 'virino' hypothesis was

Table 1.8: Possible nature of the scrapie agent

Prion	Virino
Nature	
Protein-only infectious agent (biocatalytic protein)	Protein/small nucleic acid infectious agent
Evidence	
Chemical inactivation Ultraviolet inactivation Protease inactivation	Stability of clumped virus
Infectious nature of purified PrP protein	Co-purification of nucleic acids
Failure to identify specific nucleic acid	Strain variability
	10–12 nm virus-like particles associated with fibrils

proposed suggesting that scrapie infectivity involves a small (undetected) nucleic acid associated with the PrP protein. There is also some evidence to support the virino hypothesis (see *Table 1.8*). This includes the finding that nucleic acids may co-purify with infectivity, and an observation of very small (10–12 nm) virus-like particles associated with fibrils in hamsters. However, this is balanced by evidence that highly purified protein can be infectious and, since even a 10–12 nm 'virus' could be expected to have a molecular weight of about 750 kDa, such infectivity would not be expected to co-purify with a 30-kDa protein.

A 'biocatalytic' model which could explain how a protein causes disease was proposed. In the original such model, PrP^{Sc} (the disease-associated form) converts the normal (PrP^{C}) form into PrP^{Sc}, causing disease by the loss of normal PrP^{C} function. Unfortunately for this idea, transgenic mice completely lacking the PrP^{C} gene appear normal. Other studies have shown some loss of Purkinje (motor control) cells in the brain with aging and (possibly) some differences in transmission of neural impulses, but these effects are much less severe than mouse scrapie. However, it is now known that conversion of PrP^{C} to PrP^{Sc} can be mediated by PrP^{Sc}, and this proves the core of the biocatalytic hypothesis. It may be that the refolded PrP^{Sc} form exerts a direct pathogenic effect, possibly by exposing internal regions of the protein which have been linked to neuronal apoptosis.

Although no conclusive proof of the nature of the infectious agent is currently available, the balance of evidence at present favors the prion hypothesis.

Interest in prion disease has been increased greatly by the appearance of bovine spongiform encephalopathy (BSE), or 'mad cow disease'. Since its first appearance in the mid-1980s in the UK, hundreds of thousands of cases have been reported, mostly in the UK, but also in a number of other countries in Europe and elsewhere. The disease is characterized by loss of motor control and a classic spongiform encephalopathy in the afflicted animal. At first it was thought that the disease had arisen by transfer of scrapie in sheep-derived cattle feed, but, more recently, it has been suggested that the disease arose in the UK beef herd and was transmitted by cow-derived cattle feed. While this removes some concern about a 'super-scrapie' that can jump between species, it leaves the concerns surrounding an entirely novel disease (see below). The feeding of animal protein to ruminant animals is widespread, but is now being modified in the light of the events surrounding BSE. It should be noted that while some countries are 'free' of BSE, related conditions exist, notably 'downer cow syndrome' in the United States, although this is a much less well defined syndrome. Downer cows may lack spongiform pathology, but material derived from them reputedly can induce such disease in mink.

While there is no epidemiological link between scrapie and CJD, a novel form of CJD has been identified in humans, and is referred to as 'new variant' CJD (nvCJD). It is characterized by onset at an earlier age than 'classical' CJD, and by a significantly different pathology. The

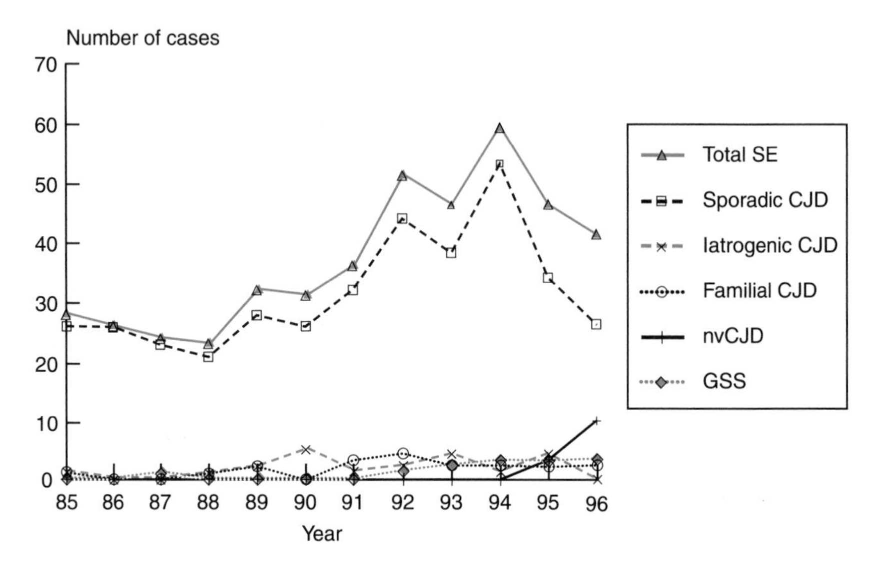

Figure 1.31: Spongiform encephalopathies (UK).

current evidence suggests that nvCJD represents a human form of BSE, and that this may have arisen from humans eating BSE-contaminated tissue. Despite all the concern and the huge economic costs resulting from BSE, the relative numbers of nvCJD are limited (*Figure 1.31*), with a total of 21 confirmed cases (20 in the UK and one in France) and two suspected cases by September 1997. The concern is that we are only seeing the first few cases of the 'bell curve' of a major epidemic, but the evidence to support this is inconclusive. In particular, the lack of an elevated rate of nvCJD in slaughterhouse workers, who would be exposed to the highest levels of the agent (and, from animal studies with scrapie, would be expected to show a shorter incubation period for the disease as a result), suggests that transmission to humans may be rare. It is known that genetic factors influence susceptibility to nvCJD. All cases identified to date have been homozygous for methionine at residue 129 of PrP. This particular genotype is only present in 39% of the UK population. It is possible that other (unknown) genetic factors may also be required for the disease to become established, further restricting the susceptible population, but this is not yet known. In addition, there has been an overall decline in the number of cases of CJD (and also in the total number of prion diseases) in the UK from a 1994 peak, despite intensive surveillance efforts. These data may indicate that the outbreak of nvCJD will remain limited. Only time will provide the answers to this question.

Further reading

Arrand, J.A. and Harper, D.R. (1998) *Viruses and Human Cancer*. BIOS Scientific Publishers, Oxford.

Butler, P.J. and Klug, A. (1978) The assembly of a virus. *Sci. Am.*, **239,** 62–69.

D'Halluin, J.C. (1995) Virus assembly. *Curr. Top. Microbiol. Immunol.*, **199,** 47–66.

Fields, B.N., Knipe, D.M. and Howley P.M. (1996) *Virology*, 3rd Edn. Raven Press, New York.

Garcia-Blanco, M.A. and Cullen B.R. (1991) Molecular basis of latency in pathogenic human viruses. *Science*, **254,** 815–820.

Haywood, A.M. (1994) Virus receptors: binding, adhesion strengthening, and changes in viral structure. *J. Virol.*, **68,** 1–5.

Hogle, J.M., Chow, M. and Filman, D.J. (1987) The structure of poliovirus. *Sci. Am.*, **256**, 42–49.

Ironside, J.W. (1996) Human prion diseases. *J. Neural Transm.*, Suppl. **47,** 231–246.

Kaplan, M.M. and Webster, R.G. (1977) The epidemiology of influenza. *Sci. Am.*, **237,** 88–106.

MacDonald, F. and Ford, C.H.J. (1997) *Molecular Biology of Cancer*. BIOS Scientific Publishers, Oxford.

Madeley, C.R. and Field, A.M. (1988) *Virus Morphology*, 2nd Edn. Churchill Livingstone, Edinburgh.

Minson, A., Neil, J. and McCrae, M. (1994). *Viruses and Cancer: 51st Symposium of the Society for General Microbiology*. Cambridge University Press, Cambridge.

Murphy, F.A., Fauquet, C.M., Bishop, D.H.L., Ghabrial, S.A., Jarvis,, A.W., Martelli, G.P., Mayo, M.A. and Summers, M.D. (Eds) (1995) Virus Taxonomy: Classification and Nomenclature of Viruses: Sixth Report of the International Committee on the Taxonomy of Viruses. *Arch. Virol.*, Suppl. 10.

Nathanson, N., Wilesmith, J. and Griot, C. (1996) Bovine spongiform encephalopathy (BSE): causes and consequences of a common source epidemic. *Am. J. Epidemiol.*, **145,** 959–969.

Oldstone, M.B. and Rall, G.F. (1993) Mechanism and consequence of viral persistence in cells of the immune system and neurons. *Intervirology*, **35,** 116–121.

Reanney, D. (1984) The molecular evolution of viruses. In *The Microbe 1984, 1: Viruses* (B.W.J. Mahy and J.R. Pattison, Eds). Cambridge University Press, Cambridge, pp. 175–196.

Sanger, H.L. (1984) Minimal infectious agent – the viroids. In *The Microbe 1984, 1: Viruses* (B.W.J. Mahy and J.R. Pattison, Eds). Cambridge University Press, Cambridge, pp. 281–334.

Somerville, R.A., Bendheim, P.E. and Bolton, D.C. (1991) Debate: the transmissible agent containing scrapie must contain more than protein. *Rev. Med. Virol.*, **1**, 131–144.

Electronic resources

For specific sources, use of a search engine such as Alta Vista or Webcrawler will provide links to relevant sites and, since URL addresses change frequently, may be more up to date than the links provided below. Some useful examples are listed below, and provide a variety of links to related material:

'All the Virology on the WWW', general virology site
http://www.tulane.edu/~dmsander/garryfavweb.html
(mirror sites exist outside US)

Bionet virology newsgroup
News:bionet.virology

'Index Virum', from the 6th report of the International Committee on the Taxonomy of Viruses
http://life.anu.edu.au/viruses/Ictv/index.html

Institute for Molecular Virology, University of Wisconsin Madison, general virology site
http://www.bocklabs.wisc.edu/Welcome.html

Medweb: Infectious Diseases, many links
http://www.gen.emory.edu/medweb/medweb.id.html

The Official Mad Cow Disease Home Page, BSE and prion disease information
http://www.mad-cow.org

Chapter 2

Viral interactions with the immune system

The immune response to virus infection is of paramount importance, and this is amply demonstrated by the severe nature of virus disease in children who lack normal immunity due to inborn genetic errors. The inhibition of virus replication by the immune system limits the spread of infection and moderates the course of disease. Without an immune response mounted by the host, virus infection would be fatal, as seen in the killing of cultured cells by viruses.

The human immune response to virus infection consists of the non-specific and the adaptive responses, and can be divided into three parts. These are:

(1) the non-specific (innate) immune response;
(2) the cell-mediated immune response (CMI);
(3) the antibody response.

2.1 The non-specific immune response

This involves cytokines and natural killer (NK) cells, as well as the complement response.

Cytokines are regulatory proteins produced by a wide range of cells. They act over short distances, and are involved in many aspects of virus infection. More than 40 cytokines have been identified, although many of these are not thoroughly characterized. One of the most significant groups of cytokines in the context of virus infection are the interferons. There are three basic types of interferon, listed in *Table 2.1*, as well as some minor species. Interferons are produced in response to a range of stimuli (including many arising from virus infection), induce an antiviral state in the infected cell, and are released to produce similar effects in nearby cells. The interferon response is very rapid, appearing within hours of viral infection, and is an important element of pre-specific immunity. Broadly,

Table 2.1: Properties of interferons

Interferon	Producer cells	Physical properties[a]	Inducers
α-Interferon[b,c]	Leukocytes (fibroblasts)	Acid-stable, non-glycosylated protein	dsRNA Virus infection
β-Interferon[c]	Fibroblasts (epithelial cells)	Acid-stable glycoprotein	dsRNA Virus infection Bacterial components Cytokines (TNF, IL-1)
γ-Interferon[d]	T cells NK cells	Acid-labile glycoprotein	Antigens Mitogens Cytokines (IL-2)

[a]All interferons are proteins of approximately 20 kDa.
[b]Human α-interferon is produced from at least 15 closely related genes, β- and γ-interferons from single copy genes.
[c]α- and β-interferons may be referred to as type I interferons, intermediate forms exist.
[d]γ-interferon may be referred to as type II or immune interferon.

interferons make cells resistant to virus infection by a wide range of effects, only some of which have been explained. Interferon-mediated effects are summarized in *Tables 2.2* and *2.3*. Other cytokines important in the response to virus infection include tumor necrosis factors (TNFs) and interleukins (ILs), which (among a vast range of other effects) stimulate lymphocyte proliferation. The cellular element of the pre-specific response is provided by NK cells. These are cytotoxic lymphocytes which lack a T-cell receptor (TCR). They are targeted to the site of virus infection by cytokines before specific cytotoxic T cells (CTLs) are produced. NK cells recognize target cells to which antibody is bound, referred to as antibody-

Table 2.2: Interferon: the antiviral state

Synthesis of:	Ribonuclease L Protein kinase PKR/P1 2′-5′ oligo-A synthetase
Activation of:	2′-5′ oligo-A synthetase by dsRNA Ribonuclease L by 2′-5′ oligo-A Protein kinase PKR/P1 by dsRNA M_x gene transcription (mice; α, β-interferon) Synthesis of many other cellular proteins
Inactivation of:	Eukaryotic initiation factor 2 (eIF2) by protein kinase Down-regulation of cellular genes and oncogenes
Effects:	Non-specific RNA degradation (ribonuclease L) Non-specific inhibition of protein synthesis (due to eIF2 inactivation) Effects on transcription (PKR/P1) Inhibition of viral penetration/uncoating (may involve membrane alterations) Inhibition of primary viral transcription (influenza, via M_x gene product) (other viruses, unknown mechanisms) Inhibition of retroviral cell transformation (may involve membrane alterations) ↓ INHIBITION OF VIRAL REPLICATION Inhibition of tumor cell growth

Table 2.3: Interferon: extracellular effects

Enhancement of:	Inhibition of:
MHC-I presentation	Viral replication (by induction of antiviral state)
MHC-II presentation (β,γ-interferon)	Cell division
T and NK cell cytotoxicity	Antibody production (α-interferon)
Antibody production (γ-interferon)	
Macrophage activation (γ-interferon)	

dependent cellular cytotoxicity (ADCC, see Section 2.3). Other mechanisms by which NK cells identify target cells are poorly defined, but a significant factor appears to be that they recognize 'lack of self', and will not lyse cells expressing autologous major histocompatibility complex class I (MHC-I) molecules. NK cells release perforin which disrupts the integrity of the target cell membranes, as well as inducing apoptosis in the target cell by the release of granzymes (serine proteinases). Usually, NK cells are only weakly cytotoxic. However, in the presence of α- or β-interferon (levels of which are raised by virus infection) or interleukin 12, NK cell activity is enhanced. Viral proteins may also enhance NK function, either directly or by inhibiting MHC-I expression. A range of viruses have such an inhibitory effect, apparently as a way of limiting the cell-mediated immune response (see Sections 2.2 and 2.6). However, the effect of NK cell activity is moderated by a second consequence of virus-induced interferon production: up-regulation of class I MHC gene transcription, which may serve to balance the stimulation of NK cell activity by interferons. NK cell activity declines within a few days of virus infection as the specific immune responses appear in the host, and it has been suggested that they represent an evolutionary 'hangover' from before the development of the adaptive T-cell response. Other pre-adaptive lymphocytes exist, notably the γ/δ cells, which appear to be involved in immunity at epithelial surfaces and have been implicated recently in controlling herpes simplex encephalitis.

A further non-specific system is the ability of the complement system of serum proteins to become activated by the surface molecules of some pathogens. This is referred to as the 'alternative pathway' of complement activation, since it is not dependent on specific antibody, unlike the 'classical pathway' discussed in Section 2.3. Complement activation then results in the formation of the complement 'membrane attack complex' which can kill a bacterium or destroy a virus. Unusually, the surface glycoprotein of HIV appears to be able to activate both complement pathways directly. Other functions of the complement system are discussed in Section 2.3.

2.2 The cell-mediated immune response

This is summarized in *Figure 2.1*. As with the antibody response, T cells respond to specific antigens, and some memory T cells will persist even

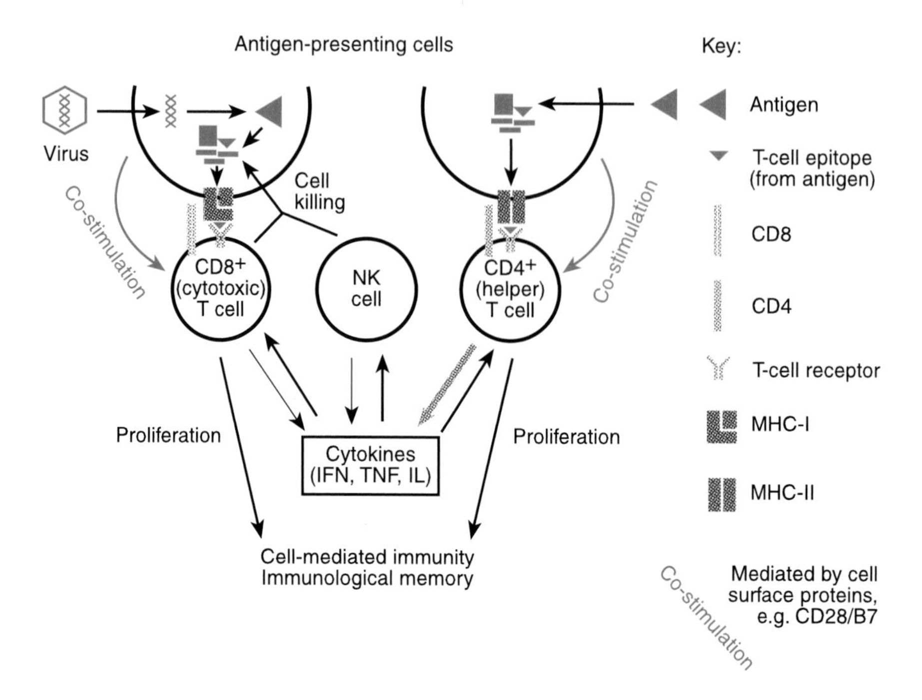

Figure 2.1: Basic features of the cell-mediated immune response.

after infection resolves, allowing a more rapid and directed response to later challenge with the same antigen. It is currently not entirely clear whether this immunological memory is provided by dedicated memory T cells in all cases, since antigenic restimulation by low level infections or persisting antigen may also play a role. Unlike B cells producing antibodies, T cells respond to peptide antigens composed of 8–10 adjacent amino acids in the MHC-I pathway, or at least 13 adjacent amino acids in the MHC-II pathway. These peptides are produced by cleavage of the antigenic protein, and are presented on the surface of cells in association with specific molecules. Antigens taken up from the extracellular space by cells of the immune system are processed in degradative endosomes and presented in association with MHC-II molecules, while antigens produced in or introduced into the cytoplasm are processed there and presented in association with MHC-I molecules (*Figure 2.2*). The nature of the presenting molecule determines the effect of antigenic presentation as discussed in Sections 2.2.1 and 2.2.2.

The range of epitopes that can be presented for recognition by T cells is huge, and the necessary variability in the TCR is produced by genetic recombination processes similar to those involved in producing antibodies (see Section 2.3), to which the TCR is closely related. Multiple T-cell epitopes may be present on one protein, and are distinct from the B-cell epitopes which induce an antibody response. There are two main pathways of antigen presentation to T cells: the MHC-I and the MHC-II pathways.

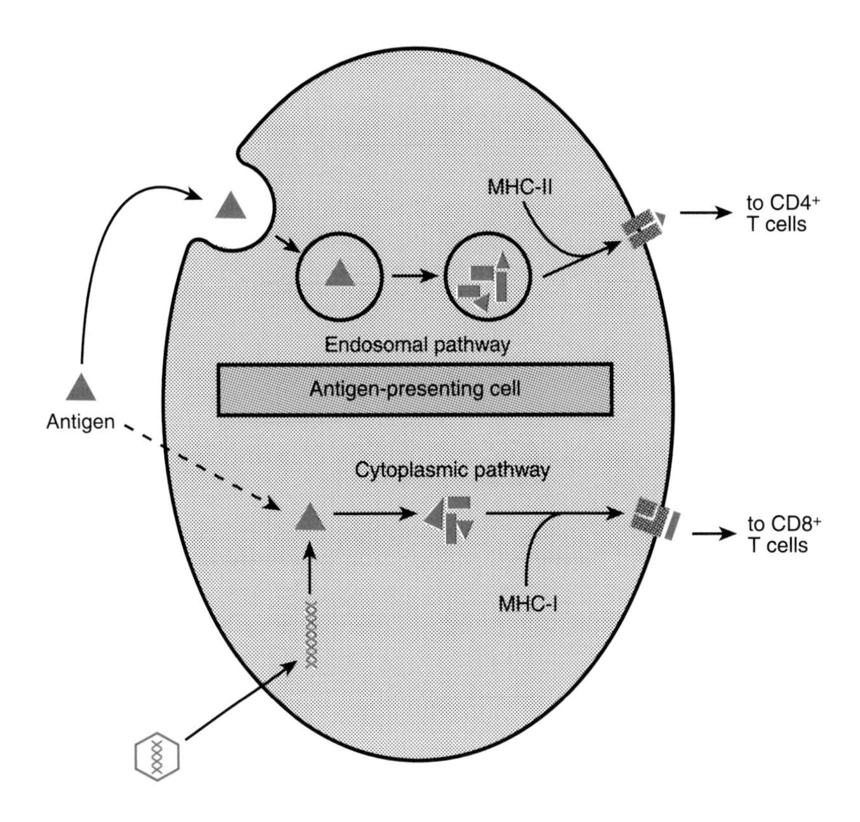

Figure 2.2: Presentation of T-cell epitopes; endosomal and cytoplasmic pathways.

The MHC-I pathway predominantly activates cytotoxic T cells, while the MHC-II pathway mainly activates helper T cells that facilitate the immune response (although there are exceptions in both cases). The two pathways are discussed in more detail below.

2.2.1 The MHC-I pathway

Peptides of approximately 9–13 amino acids are produced from antigenic proteins by digestion by the proteasome, a proteinase complex in the cytoplasm of the cell. The peptides are then passed into the endoplasmic reticulum by the TAP transporter protein and, after further trimming to 8–10 amino acids in length, they become associated with compatible TAP-bound MHC-I molecules. Until recently, it was thought that antigens had to be synthesized within the cell in order to be entered into this pathway, but it is now clear that exogenous proteins which are entered into the cytoplasm by osmotic shock, fusogenic liposomes or some vaccine adjuvants (see Section 3.2) can also be presented via the MHC-I pathway. Peptides must fit certain structural requirements, and are inserted into a deep groove on the surface of MHC-I molecules (see *Figure 2.3*) within the

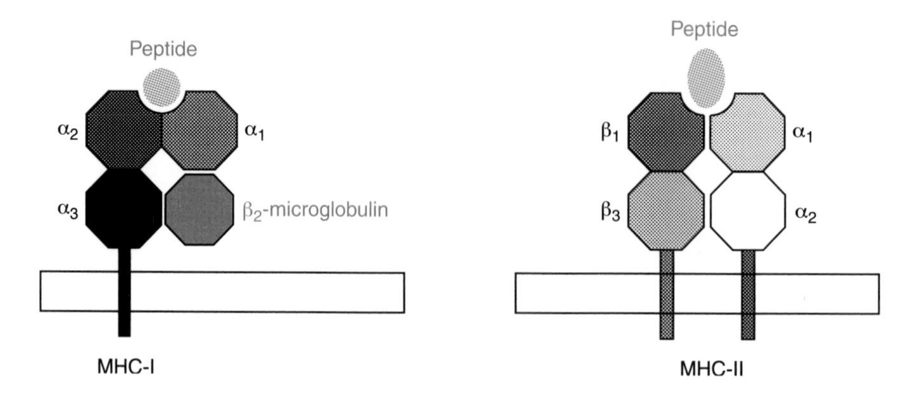

Figure 2.3: Structure of MHC molecules.

endoplasmic reticulum of an infected cell, and the antigen–MHC-I complexes are presented on the cell surface. Binding of these complexes to the TCR stimulates the T cells via the associated CD3 proteins and ζ (zeta) chains, which transduce a signal across the membrane. Other adhesion proteins are also involved in T cell binding, notably CD2 and LFA-1 on the T cell binding to LFA-3 (CD58) and ICAM-1 or -2 on the antigen-presenting cell. Additionally, ICAM-3 on the T cell binds to LFA-1 on the antigen-presenting cell. Since MHC-I proteins are present on almost all cells and presentation by the MHC-I pathway is, in most cases, limited to proteins actually synthesized within the presenting cell, the MHC-I pathway is a method for cells to call in a T-cell response if they become infected. MHC-I molecules are produced from three highly variable genes on each human chromosome 6, giving up to six different forms with varying binding characteristics. This variation allows presentation of a broader range of peptides. Presentation of antigen with MHC-I molecules will not activate all T cells, but specifically those with TCRs matching the expressed MHC-I molecule (MHC restriction). Successful interaction activates T cells with the CD8 surface marker protein ($CD8^+$ T cells), which are mostly CTLs. These then kill the presenting cell by mechanisms that appear to be similar to those used by NK cells, involving cytokine synthesis, perforin release and the induction of apoptosis by granzyme proteinases (see Section 2.4). CTLs, by killing virus-infected cells, prevent the spread of virus and protect the human host from further harm in most instances. However, there are situations where T cells can cause direct harm, such as in intracerebral infection of mice with the arenavirus, lymphocytic choriomeningitis virus (LCMV). In the absence of a competent T-cell response, a persistent, non-lethal LCMV infection is established. In the presence of competent T cells, intracerebral LCMV infection is highly destructive, and kills the mouse. In human disease, T cell-mediated cytotoxicity is believed to be significant in viral hepatitis. The cytotoxic T-cell response is very powerful, and plays an important role in controlling virus infection. Despite this, it can be very damaging and of necessity is tightly controlled, for example by the need for co-stimulation (Section 2.5), the absence of which can lead to tolerance of the expressed antigen, and by the presence of suppressor T cells, which moderate the activity of other T cells.

2.2.2 The MHC-II pathway

Unlike the MHC-I pathway, processed antigens presented by the MHC-II pathway do not need to be synthesized within the presenting cell or entered into the cytoplasm by specialized systems. Rather, these are usually proteins taken up by specialized cells and processed within degradative endosomes. The peptides produced in this way are presented on the cell surface in association with the MHC-II proteins. The size of peptide presented by MHC-II molecules is both larger and less rigid than with MHC-I molecules since the MHC-II-binding groove is open-ended. The peptide is usually at least 13 amino acids in length, but can be far longer. However, it will usually be trimmed to a maximum of 17 amino acids after binding to MHC-II. As with MHC-I, MHC-II proteins are produced from multiple, highly variable genes. In the case of MHC-II, four genes on each human chromosome 6 give eight possible variants. Peptide–MHC-II complexes are recognized by T cells with the CD4 surface marker protein ($CD4^+$ T cells) via their TCR–T3 complex. As with the MHC-I pathway, other adhesion proteins are also involved [notably CD2/LFA-3(CD58), LFA-1/ICAM-1 or -2 and ICAM-3/LFA-1, as noted above]. The MHC-II proteins are generally present only on a limited range of antigen-presenting cells associated with the immune system, although some other cell types appear to express them, including keratinocytes in the skin. Cells of the immune system presenting antigens by the MHC-II pathway are performing their usual task of picking up foreign proteins for presentation to other elements of the immune system, rather than identifying themselves as infected, so it would be inappropriate for them to be killed. Instead, the main type of T cell activated by this pathway is helper T cells. In response to antigenic stimulation, these cells proliferate and secrete cytokines which activate both the antigen-presenting cell and other cells of the immune system. Helper T cells are vital to the activation of many components of the immune system, including NK cells (by IL-2 and IFN-γ), CTLs (IL-2, IL-4, IL-6) and B cells (IL-4, IL5, IL-6). γ-Interferon produced by helper T cells also up-regulates MHC-II expression on cells, including some types of cell which do not generally express these proteins. While it is possible to list the major activities of the cytokines produced by these cells, it is important to note that each has multiple functions, and that the interactions of cytokines within the immune system are a hugely complex area of study.

Helper T cells exist in two broad classes, T_H1 cells, which activate macrophages and other cellular elements in the cytotoxic (cellular) response, and T_H2 cells, which activate B cells and the humoral (antibody) response. Activation of a T_H1 or T_H2 often leads to that element of the response becoming dominant, so that either cellular or humoral responses predominate. The cytokines produced by T_H1 or T_H2 responses differ, and are often used to determine the nature of the response, with, for example, the presence of IL-2, IL-12 and IFN-γ indicating a T_H1 response, while IL-4, IL-5 and IL-10 indicate a T_H2 response.

Cytokines produced by T cells are also involved in delayed-type hypersensitivity. These cytokines increase the permeability of capillaries at the site of infection, helping other elements of the cellular immune response to get to where they are needed, along with serum proteins such as fibrin. These effects produce a characteristic local induration of the site of infection, initially observed in the local reaction to tuberculosis (a bacterial disease), which is referred to as delayed-type hypersensitivity. This response is also thought to be important in mediating local reactions to viral infections, such as the formation of vesicular lesions.

2.3 The serological immune response

This is mediated by antibodies, which are glycoproteins with the ability to bind to specific molecular structures on their target antigen. They are produced by plasma cells. These are large, highly specialized cells produced by differentiation of B cells, which are adapted for very high levels of protein synthesis. The process of B-cell activation and antibody synthesis is shown in *Figure 2.4*. Antigens are recognized by the B cell's own immunoglobulin molecules, which are present on the cell surface. Thus, the immunoglobulin molecule produced by a particular B cell serves as the receptor for its target antigen. After binding, the antigen is internalized and cleaved into peptides which are transported back to the cell surface complexed with class II MHC molecules. This and other MHC-II-presenting cells stimulate helper T cells to produce cytokines (including IL-4, IL-5 and IL-6) and membrane signals (mainly CD40 ligand produced by the T cell combining with CD40 on the B cell) which, combined with the effects of antigen binding, induce B-cell proliferation and differentiation. This system is summarized as the 'clonal selection hypothesis'. Each B cell encodes a specific receptor. When the receptor is engaged by antigen, the B cell divides and forms a clone of daughter cells with identical receptors, a process modulated by MHC presentation, T-cell effects and cytokines. Even though B cells able to produce a vast range of antibodies are produced, essentially at random, only B cells which bind antigen are stimulated to proliferate, avoiding the production of 'useless' antibodies (*Figure 2.5*).

The basic structure of immunoglobulin G is shown in *Figure 2.6*. Antibodies have a mostly constant structure, tipped by highly variable

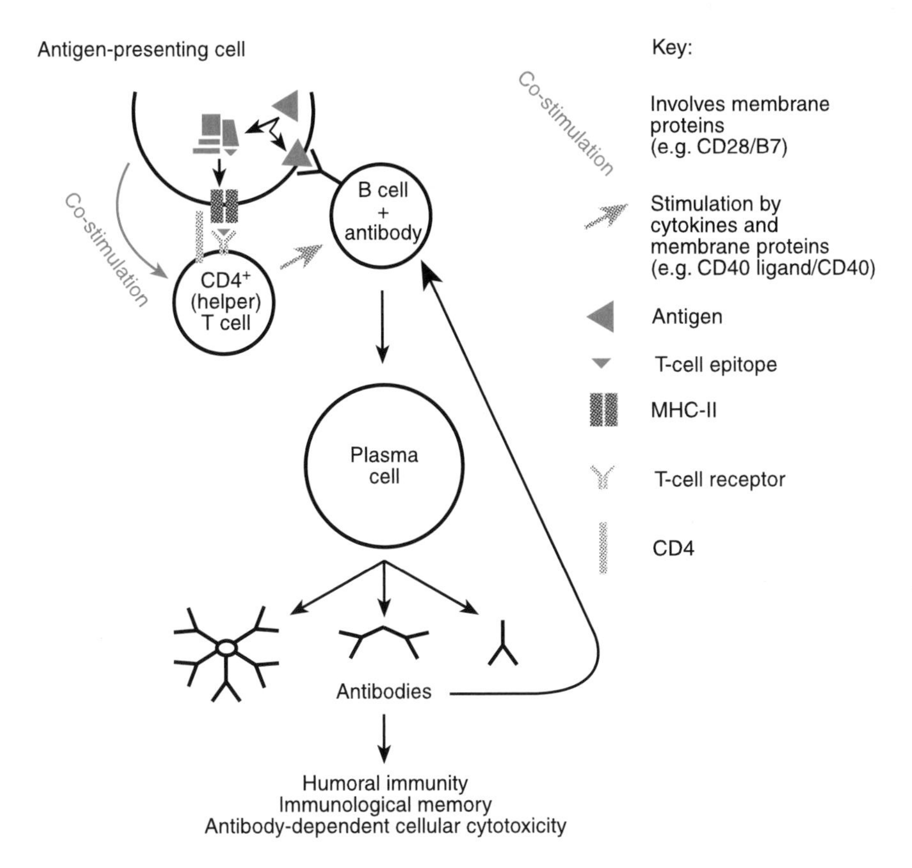

Figure 2.4: Basic features of the antibody response.

regions which enable the antibody to bind to a specific antigen. The immunoglobulin response is extremely complex because of the enormous diversity within the system. Within each individual there may be as many as 10^9 different antibody molecules, each specific for a different antigen. Antibodies are predominantly of constant structure. Even with the variable regions (see *Figure 2.6*), variation is concentrated in three hypervariable regions in each light chain and in each heavy chain. These form paired regions in the heavy and light chains which are referred to as the complementarity-determining regions; CDR1, CDR2 and CDR3. The CDRs combine to form the unique antigen-binding site. The diversity of antigen binding arises from the process by which antibodies are formed. Rather than each being coded for by a specific gene (which would require rather too much DNA to be practical and would not allow the flexibility to cope with novel antigens), the variable regions are encoded by a wide range of small genetic elements. During B-cell maturation, multiple variable elements are spliced together by special enzymes which induce further variation by altering bases at the splice site. The net effect of all this is a huge range of possible structures for the variable regions which

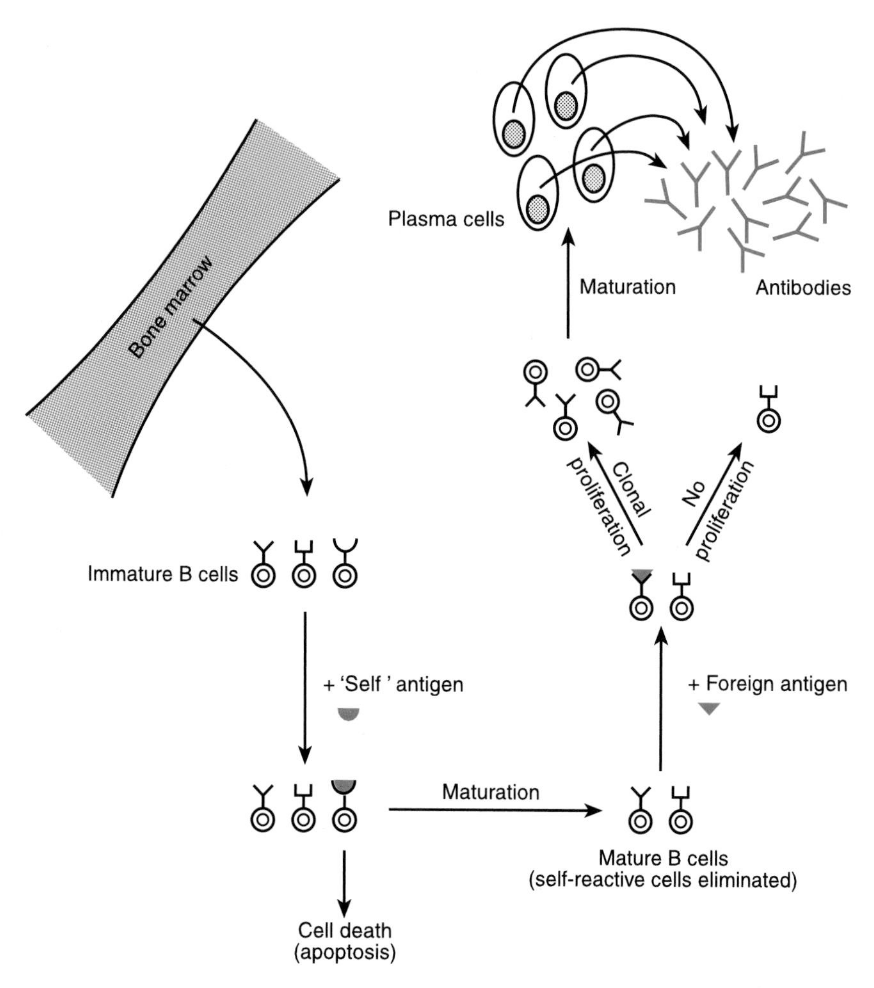

Figure 2.5: B-cell maturation and proliferation. (Similar selection by cell death and clonal proliferation occurs during T-cell maturation.) Co-stimulation events not shown (see *Figure 2.3*).

are produced essentially at random, so that an antibody can be produced to almost any antigen, including totally novel structures. Cells that produce antibodies recognizing 'self' molecules of the host organism are destroyed during maturation, and the remainder are released to await possible stimulation and clonal selection (*Figure 2.5*).

Unlike the presentation of processed peptides to T cells, B cells recognize antigens in their native form. For viruses, this is frequently the outer proteins of the viral particle, such as the glycoproteins of enveloped viruses (which are often also present on the surface of infected cells). Proteins that generate an immune response are referred to as 'immunogenic'. In general, an immunogenic protein contains several distinct sites which can be bound by an immunoglobulin (IgM or IgD) receptor on

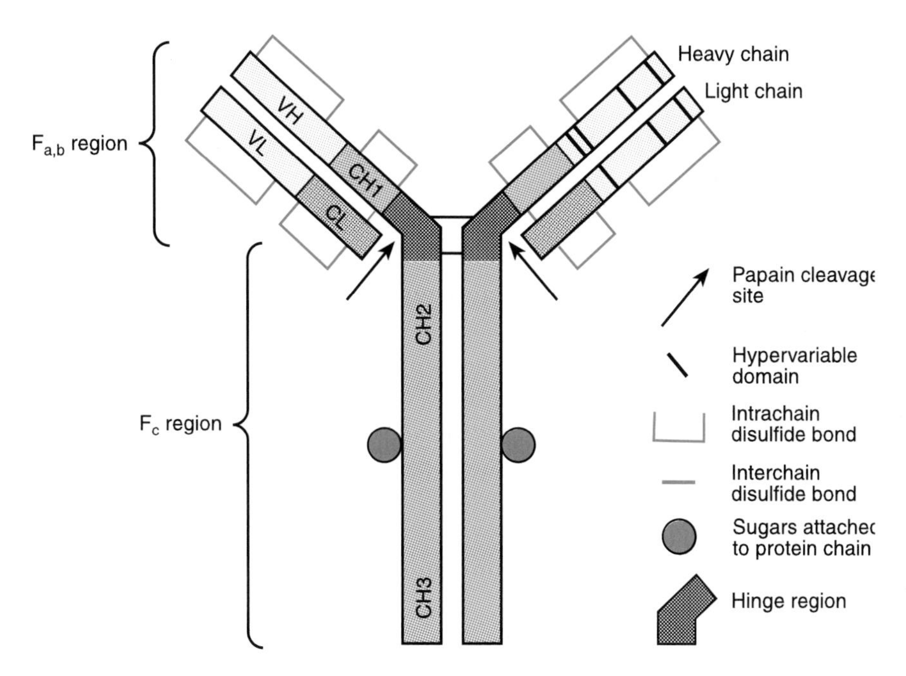

Figure 2.6: The structure of immunoglobulin G. CH, constant region of heavy chain; VH, variable region of heavy chain; CL, constant region of light chain; VL, variable region of light chain. Other immunoglobulin isotypes have additional carbohydrate, and both IgM and IgE lack the hinge region and have an additional CH region.

different B cells. Each of these antigenic sites is called a determinant or an epitope (see Section 2.7). Epitopes are often defined with the use of monoclonal antibodies (see Section 2.8.2), which bind to a single specific epitope on a protein. The use of different monoclonal antibodies can show the presence of different epitopes, or show that one highly immunogenic region of the protein induces the production of multiple antibodies (which may then interfere with each other's binding). Such highly immunogenic regions are useful in the design of vaccines (see Chapter 3). B-cell epitopes may consist of a short stretch of amino acids (a linear epitope) or, more often, a cluster of amino acids from different regions of the antigenic protein which are brought close together only when the molecule acquires its correct three-dimensional structure (a discontinuous or disperse epitope, also referred to as a conformational epitope, discussed in Section 2.7). Denaturation of a protein with detergents or other reagents (as is used in Western blotting) may destroy conformational epitopes, but linear epitopes are resistant to denaturation. It is often the case that biologically important viral epitopes are conformational, and this includes many of those involved in neutralization of enveloped viruses. This limits the use of denaturing assay systems in examining the humoral immune response. Relatively non-denaturing systems such as radioimmune precipitation can be used to examine conformational epitopes. Specific recognition of a viral protein by the antibody, whether on the virus particle or the surface of an infected cell, can result in direct inhibition of viral infectivity (neutralization) if an essential viral function is prevented by antibody binding. Protein–protein interactions may also mean that antibody binding has effects inside the virus. Alternatively, the bound antibody can target other elements of the immune system. These include:

- phagocytic cells which engulf and digest antibody-coated virus;
- ADCC, where NK cells are targeted by Fc receptors expressed on their membranes to kill cells to which antibody is bound;
- the complement system (by the antibody-dependent 'classical pathway'), which has multiple functions:
  (i) enhancing neutralization by masking or agglutinating virus;
  (ii) stimulating uptake by phagocytes;
  (iii) forming the membrane attack complex which is able to create lethal pores in membranes where appropriate antibody is attached;
  (iv) mediating inflammation and phagocyte recruitment.

The humoral immune response involves several different types of antibody, summarized in *Table 2.4*. The same plasma cell can produce different types of antibody by 'switching' the constant regions attached to the antibody produced. The antibody response follows a characteristic profile, with IgM appearing first, followed by IgA and IgG. These are the major immunoglobulin isotypes. IgA and IgG can be subdivided further into subtypes called IgA1, IgA2, IgG1, IgG2, IgG3 and IgG4. IgG is the major element of humoral immunity and will remain present in the blood at low but detectable levels long after the resolution of infection, reflecting the presence of 'memory' B cells, which allow a rapid antibody response to any further challenge with the specific antigen. Circulating IgM is a pentameric molecule containing a small joining protein (the J chain), although a monomeric form is present on the surface of B cells where it acts as a receptor. IgM is produced in more primitive organisms than is IgG and may be most significant in bacterial infections. IgM is produced rapidly following infection, but does not then persist. Much of the IgA produced is dimeric (linked by J chain) and is targeted to and secreted on to mucosal surfaces, where it forms a 'first line of defense'. Serum and mucosal IgA have distinct functions and distinct structures. Like IgM, IgA does not persist. IgE and IgD are produced in smaller amounts; IgE is primarily concerned with the hypersensitivity and inflammatory responses, while IgD is a receptor on mature B-lymphocyte membranes.

The isotype of antibody produced by B cells from one clone is not fixed, but rather alters as the immune response proceeds. IgM is produced first, and isotype switching is controlled by cytokines and helper T cells (and is mediated by CD40 binding). A cell that initially was producing IgM may be induced to produce IgG, IgA or IgE, but while the class of antibody changes, this involves the CH regions only. As a result, the

Table 2.4: Functions and properties of immunoglobulins

	IgG	IgM	IgA	IgD	IgE
Major functions	Extravascular immunity Anamnestic responses	Agglutination Defense against bacteremia Primary immune response	Secretory form mediates immunity on mucosal surfaces	Present on B-cell surface	Anaphylaxis Allergy
Molecular weight (kDa)	146–165	970 (pentameric)	160 (secretory form is dimeric, 390)	184	188
Subclasses	4	1	2 + secretory	1	1
Carbohydrate content (%)	3	8	12	13	12
Chain types:					
Heavy	γ_{1-4}	μ	α_{1-2}	δ	ε
Light		All immunoglobulins have both κ and λ light chains			
Domains in heavy chain	4	5	4	4	5
Disulfide bonds between heavy chains	2–11	2	1	1	2
Mean concentration in serum (mg ml^{-1})	13.5	1.5	3.5	0.03	0.00005
Half-life in serum (days)	7–21	10	6	3	2
Complement activation:					
Classical	– to + + +	+ + + +	–	–	–
Alternative	–	–	– to +	–	+
Placental transfer	+ + +	–	–	–	–
Antibody-dependent cellular cytotoxicity (ADCC)	– to + +	–	–	–	–

binding specificity of the antibody is unaltered, but the consequences of that binding are changed since the CH regions mediate the effects of bound antibody on other elements of the immune system (see *Table 2.4*). However, the binding properties of the antibody produced do alter during the immune response. Due to a high rate of mutation in antibody-producing cells, variations in antibody structure, and thus in affinity of antigen binding occur. Clearly, those cells with higher affinity antibody on their surface will bind more strongly to antigen, and thus be more strongly induced to proliferate. This will have the effect of reducing antigen binding by cells producing unaltered antibody and those where mutation has resulted in a decrease in affinity. Such cells will then not be induced to proliferate, the plasma cells producing the antibodies will not be replaced, and antibody affinity will increase during the initial phases of the immune response.

The affinity of antibodies in the blood can be measured, and the presence of low affinity antibodies shows that the immune response is in its early stages. In another, more commonly used diagnostic application, the humoral immune response is often used to determine whether a patient has been infected by a particular virus. Demonstration of specific IgG can show immunity resulting from a previous infection, while IgM shows that a recent infection has occurred.

An important concept in humoral immunity to virus infection is the hierarchy of the response. Although a virus has multiple antigens, the immune response is not equally distributed among these. Rather, there is a pronounced response to some ('immunodominant') antigens and a minimal response to others. At present, there are only limited ways to predict which will be the preferred antigens. Instead, the antibody responses to individual antigens are assessed and compared in sera taken from patients with typical viral infections. Over time, following viral infection, antibodies to some viral proteins will disappear while those to other antigens will remain detectable for many years. These issues must be considered when designing assays of humoral immunity to a particular virus, or producing vaccines or therapeutic antibody preparations.

The humoral immune response is very important in the control of some viral infections, notably those viruses which are present at high levels in a cell-free form (such as poliovirus), while the CMI is more significant for viruses which are usually associated with cells (such as herpesviruses). Poliovirus (see Section 1.4.5, for details of structure and replication) stimulates a humoral response consisting of IgG and IgM in the blood as well as a local (secretory IgA) response in the gut. The IgG response (which can neutralize the virus directly) then persists for the lifetime of the individual and can prevent further infection.

The above represents active humoral immunity, where antibody is synthesized by the infected individual. It is also possible to give 'passive' humoral immunity by the injection of antibody preparations (see Section 2.8).

2.4 Apoptosis

Apoptosis is a programed pathway of cell death in which the cell 'dies from within'. While it is used by the body to destroy cells once they are no longer needed or when they are considered dangerous to the organism, it is now clear that almost all cells are programed to undergo apoptosis, and that in many cases external stimuli are required to prevent apoptosis from occurring, including cytokines and a complex range of signaling mechanisms. Apoptosis can be triggered by a number of events, including many associated with virus infection, such as disruption of the cell cycle. It is also possible for apoptosis to be induced from outside the cell, and this is a major method for controlling pre-malignant cells within the body. Apoptosis also provides the main mechanism by which CTLs and NK cells kill their targets (see Section 2.2.1). Apoptosis involves multiple pathways within the cell and has a range of characteristic effects, which have been studied extensively in lymphocytes. These effects include: shrinking of the cell, clumping and breaking of the nuclear DNA (often into a characteristic 'ladder' of fragments of differing sizes), break up of the nucleus, disruption of the cell membrane and, finally, breakdown of the cell into 'apoptotic bodies', which are cleared by phagocytosis. Several cellular oncogenes are involved in apoptosis, notably the *p53* tumor suppressor gene product (see Section 1.7) which suppresses cell proliferation and can induce apoptosis. Loss or suppression of *p53* function is associated with the appearance of many cancers. There are also proteins that have an anti-apoptotic effect. For example, the proteins of the *bcl-2* family appear to be involved in preventing apoptosis. This effect may be important in establishing long-term immunological memory, but is also seen in B-cell lymphomas. External stimuli such as co-stimulatory proteins (notably β_1 integrins) may also be involved in preventing apoptosis.

Apoptosis is an important mechanism in the control of cell proliferation and the prevention of uncontrolled cell growth. However, it has become clear that it is also the major means by which 'redundant' cells are removed from the body, including many cells of the immune system. Apoptosis is also central to normal cellular development and the fine tuning of cellular differentiation.

2.5 The role of co-stimulation in the immune response

While antigen presentation in the context of MHC antigens is required to induce proliferation of T cells, other stimuli are also required. Indeed, antigenic stimulus alone can actually make the cell unresponsive to further stimulation. Such factors are important in controlling T-cell function, since T cells can be highly destructive. Binding of cell surface proteins on the T cell with those on the antigen-presenting cell appears to provide the additional stimulus required for proliferation, and this is also discussed in

Section 2.2. While a full discussion of this area is outside the scope of this book, two examples serve to illustrate the principle.

One important co-stimulatory element is the binding of CD28 on the naïve T cells to the B7 proteins which are present on the surface of antigen-presenting cells. This is required for the T cell to begin proliferating. In an elegant demonstration of the feedback controls of the immune system, a close relative of CD28, CTLA-4, is present on the activated T cells and also interacts with B7, but much more strongly than CD28. However, CTLA-4/B7 binding has a negative effect, and actually limits further proliferation, preventing uncontrolled growth.

For B cells, activation by helper T cells involves both cytokines produced by the T cell and also contact between the CD40 protein on the B cell and a specific CD40 ligand produced by the T cell. In a further example of the complex interactions that make up the immune system, CD40 ligand activates B cells, which (among other things) then express the B7 proteins involved in the activation of T cells.

Co-stimulation provides an excellent example of the interdependence and complexity that appears to be characteristic of the immune response.

2.6 Evasion of immune surveillance

The ability to evade the immune response can enhance the ability of that virus to replicate in the host and, as a result, select for viruses with this ability. This is necessarily a highly complex field, and space does not allow more than a brief review of some of the better studied systems, which are summarized in *Table 2.5*.

Many viruses interfere actively with the immune system, preventing the proper functioning of some aspect of the immune response. Some mechanisms are relatively common, presumably reflecting the importance of the mechanisms affected in controlling virus infection. Other mechanisms have only been reported for one type of virus.

One key mechanism for any but a directly cytopathic virus is likely to be inhibition of cellular apoptosis (see Section 2.4), since this form of cellular suicide is a very common consequence of virus infection. Such mechanisms have been identified for members of the *Adenoviridae, Herpesviridae* and *Poxviridae*, and are likely to be considerably more widespread. A similar mechanism is interference with the interferon-induced protein kinase that inhibits protein synthesis in infected cells which is seen with viruses from at least seven families. Another common mechanism is interference with MHC-I presentation, by a variety of methods. A lack of MHC-I will stimulate NK cell cytotoxicity, but at least one virus [cytomegalovirus (CMV) herpesvirus] produces an MHC-I analog to prevent this. Interference with cytokines or complement also appears common. In contrast, the production of an enzyme that reduces inflammation by inducing steroid hormone production appears to be

Table 2.5: Examples of immune evasion in viruses infecting humans

Active	Passive
Infection of immune cells (*Herpesviridae*, *Paramyxoviridae*, *Picornaviridae*, *Retroviridae*)	Antigenic drift (RNA viruses; *Orthomyxoviridae*, *Retroviridae*)
Interference with complement function (*Herpesviridae*, *Poxviridae*)	Antigenic shift (segmented genomes; *Orthomyxoviridae*)
Interference with MHC-I presentation (*Adenoviridae*, *Herpesviridae*, *Paramyxoviridae*, *Poxviridae*)	Molecular mimicry (*Herpesviridae*, *Paramyxoviridae*)
Inhibition of NK cells via MHC-I homolog (*Herpesviridae*)	Latency (*Herpesviridae*, *Parvoviridae*, *Retroviridae*)
Interference with MCH-II presentation (*Herpesviridae*)	Masking of virus (*Herpesviridae*, *Retroviridae*)
Interference with cytokine production or function (*Adenoviridae*, *Herpesviridae*, *Hepadnaviridae*, *Poxviridae*)	
Interference with interferon-induced protein kinase (*Adenoviridae*, *Herpesviridae*, *Orthomyxoviridae*, *Picornaviridae*, *Poxviridae*, *Reoviridae*, *Retroviridae*)	
Interference with apoptosis (*Adenoviridae*, *Herpesviridae*, *Poxviridae*)	
Interference with inflammation (*Poxviridae*)	

restricted to the *Poxviridae*, while only EBV herpesvirus is known to interfere with MHC-II presentation.

Direct interference with the immune system is taken to extremes by HIV (*Retroviridae*). This virus actually uses the CD4 molecule as its initial receptor, and infects and kills $CD4^+$ T cells. This has the effect of disabling helper T cell function, and (due to the central role of such cells) progressively destroys the immune system, producing the acquired immune deficiency syndrome (AIDS). The method used involves both virus killing of cells and other effects, including apoptosis. It appears that the HIV proteinase may be directly involved in this effect by cleaving the Bcl-2 (anti-apoptosis) protein, in direct contrast to the anti-apoptosis mechanisms observed for other viruses.

Specific examples of direct interference with the immune system are too numerous to describe, but the herpesviruses deserve mention, and indeed have been described as 'immune escape artists'. Strategies used by different members of the family include down-regulation of MHC-I (CMV), interference with peptide transport for MHC-I presentation (HSV-1), production of a cytokine synthesis inhibitor analagous to IL-10 together with interference with MHC-II presentation (EBV), and interference with apoptosis (HHV-8/KSV).

Although many viruses have evolved 'active' methods of interfering directly in the immune response, 'passive' alternatives exist. In the case of influenza virus, the surface glycoproteins undergo rapid mutation and the resulting changes in their epitopes mean that the targets which are recognized by the immune system as influenza virus proteins may not be present in a mutated virus. This is 'antigenic drift', and is common to many RNA viruses due to the high mutation rate of RNA genomes (see Section 1.4.5). Viruses (like influenza) with segmented genomes can also acquire whole new genes during mixed infections, producing rapid and dramatic changes, known as 'antigenic shift'. Antigenic shift is thought to provide the immune evasion which allows the characteristic world-wide pandemics of influenza. Antigenic drift and shift are discussed in Section 1.4.5.

Another interesting approach is molecular mimicry, where viral antigens mimic epitopes present on host proteins. Examples include proteins of measles virus and of CMV. Clearly, the production of antibodies reacting with 'self' proteins causes all sorts of problems (including a range of autoimmune diseases), and many systems exist to avoid this. Thus, the virus recruits the aid of host systems to prevent the production of specific immunity.

As discussed above, viral latency represents another way for viruses to 'hide' from the immune system. While the virus itself cannot replicate during latency, the ability to become latent allows the virus to remain present until the host's immune responses are depressed and then reactivate, allowing a productive infection. This approach is used by herpesviruses, and the resulting reactivation disease can be both severe and highly infectious, allowing the virus to spread to new hosts.

Some forms of immune evasion contain elements of both active and passive approaches. The production of Fc receptors by herpesviruses both conceals the virus by coating it in immunoglobulin, and may interfere with immunoglobulin function as a result. CMV coats itself in β_2-microglobulin, which both masks it and also interferes with MHC-I presentation since β_2-microglobulin is an essential part of the MHC-I complex. In a related fashion, HIV adsorbs regulatory proteins from the complement system that then interfere with complement function.

Many of the above systems are very limited in the advantage which they confer on the virus. However, it should be remembered that, in the infected host, even a slight advantage can be enough to allow a productive virus infection rather than one controlled by the immune response.

2.7 Epitope mapping

It has been known for a long time that not all regions of a protein are equally immunogenic. Antibodies appear to be raised against proteins in their native configuration, and thus are most commonly directed against structures present on the surface of the protein. These epitopes may consist of linear strings of amino acids (linear epitopes) or of different regions of the protein brought together by folding (discontinuous or disperse epitopes, also referred to as conformational epitopes) (*Figure 2.7*). In addition, some epitopes will depend upon post-translational processing of the protein, particularly on glycosylation, which can account for more

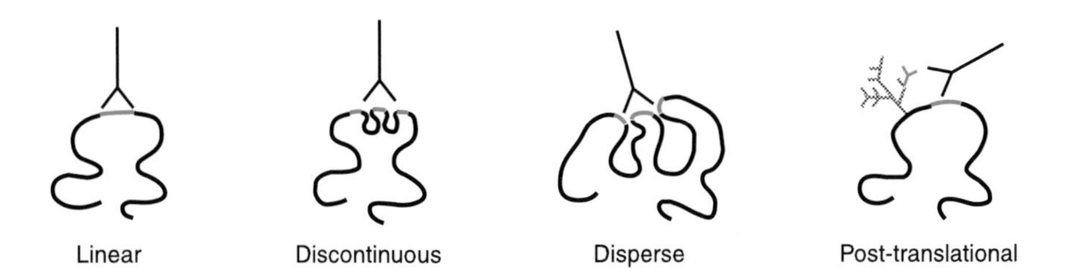

Figure 2.7: B-cell epitopes.

than half of the molecular mass of some glycoproteins. It should be noted that many systems for detecting antibodies use denatured antigen and therefore favor linear epitopes. Even binding an antigen to a solid phase will result in the loss of some discontinuous epitopes, while Western blotting is even less sensitive in detecting discontinuous epitopes due to the multiple denaturing steps involved prior to immunoreaction.

T-cell epitopes are rather different, since the processing of antigens for presentation to T cells involves their digestion into short oligopeptides. T-cell epitopes presented in conjunction with the MHC-I pathway have been identified for a range of viruses, notably influenza A and HIV. All of these consist of no more than 8–10 amino acids, and are (of course) linear epitopes. This means that 'Pepscan' mapping using short synthetic peptides is particularly suited to the identification of T-cell epitopes.

While traditional vaccines based on live or inactivated whole viruses contained multiple epitopes by default, this is not true of the many biotechnology-based vaccines under development, which often contain only fragments of viral antigens. For such vaccines it is essential to identify the immunogenic regions or epitopes within an antigen. Similarly, antigens prepared for use in diagnostic immunoassays are increasingly produced using cloning or synthetic methods. Locating epitopes can involve many different approaches, summarized in *Table 2.6*. Of these approaches, only crystallography is optimized for the detection of discontinuous or disperse epitopes. However, crystallography is laborious and technically demanding, and other approaches are frequently used. In some cases, concentrations of epitopes are apparent on particular areas of a protein. This is well illustrated by the V3 loop region of the HIV gp120 glycoprotein. This is a short (36 amino acid) highly variable region which contains a principal neutralizing determinant of gp120 (*Figure 2.8*). In addition to a neutralizing B-cell (antibody) epitope, it contains epitopes for $CD4^+$/MHC-II-dependent helper T cells and $CD8^+$/MHC-I-dependent CTLs, as well as other less well characterized determinants. Due to this concentration of immunogenic sites, the V3 loop is under intensive study for use in vaccines, although its highly variable nature makes it a difficult target.

Table 2.6: Epitope mapping

Approach	Value for epitopes			
	Linear[a]	Discontinuous[b]	Disperse[c]	Post-translational[d]
Prediction from amino acid sequence[e]	+	±	±	–
Prediction from protein structure calculations	++	+	±	±
Immunoassay of proteolytic fragments	+++	+	–	++
Immunoassay of partial clones	+++	+	–	– to ++[f]
'Pepscan' mapping using synthetic peptides	+++	–	–	–
Crystallography of antigen–antibody complex	+++	+++	+++	+++

[a]Epitope formed of amino acid residues adjacent on peptide chain (including T-cell epitopes).
[b]Epitope formed by localized protein folding, also referred to as a conformational epitope.
[c]Epitope formed from different regions of protein by complex folding, also referred to as a conformational epitope.
[d]Epitope dependent upon post-translational modifications such as glycosylation or proteolytic cleavage.
[e]Often available as part of protein analysis software.
[f]Depending upon degree of post-translational modification in expression system used (see Section 5.2).

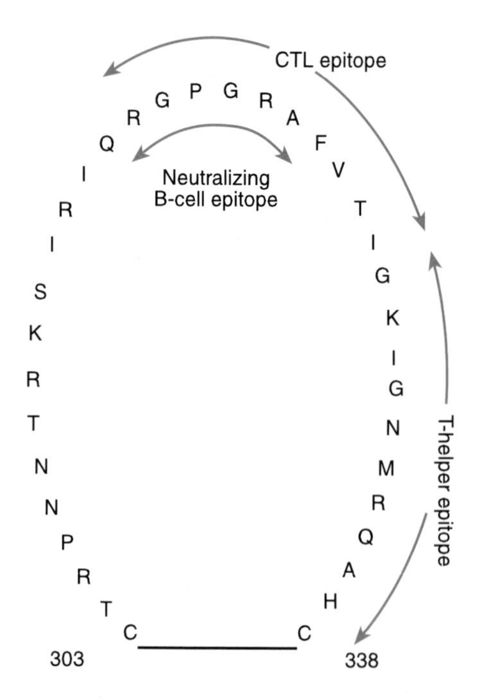

Figure 2.8: Schematic representation of V3 loop epitopes. Courtesy of Dr Linda Ebbs, British Biotechnology Ltd, Oxford, UK.

2.8 Production of specific antisera

Specific antisera have a range of applications in clinical virology, including diagnostic assays (see Section 5.2) and as therapeutic agents of passive immunity (see Section 2.9). They are also highly valuable in a wide range of research applications.

There are several approaches to the production of specific antisera. Most are based on the administration of antigen (and adjuvant) to generate an immune response to the desired antigen.

2.8.1 Monospecific antisera

Monospecific antisera at their most basic owe little to molecular technology. Following immunization, the animal (usually a rabbit or guinea pig, although larger animals may be used) is bled and the antibody response assayed. When high titers of antibody are produced, monospecific antisera are harvested. Where chickens are immunized, antibody ('IgY') may be extracted from egg yolk. Clearly the antigen used must be a single protein in order to produce a truly monospecific response. The antigen may be a purified or cloned protein, or may be a synthetic oligopeptide. As noted in the section on vaccine production, synthetic peptides will normally have to be conjugated to a carrier protein in order to be sufficiently immunogenic. Peptides often correspond to amino acid sequences predicted from nucleic acid sequence information: this represents a valuable approach to converting genome sequences into real information about proteins. The serum produced can allow identification and characterization of the corresponding protein.

While monospecific antisera are directed against a single protein, they will of course contain a wide range of different antibodies directed against multiple epitopes of the antigen. It was only with the advent of monoclonal antibody technology that it became possible to produce antibodies directed against a single epitope.

2.8.2 Monoclonal antibodies

The basic technique of monoclonal antibody production is shown in *Figure 2.9*. In the traditional approach, immunization is substantially as for monospecific antiserum production, except that mice or rats are the usual animals of choice. Antibody production is monitored but, rather than taking serum from the animal, the spleen is removed and the antibody-producing cells are immortalized by fusion with myeloma cells to produce hybridomas. Antibody-producing hybridomas are separated by dilution and culture so that each population grows from a single cell ('cloning'). As a result, it is common to use multiple antigens and derive

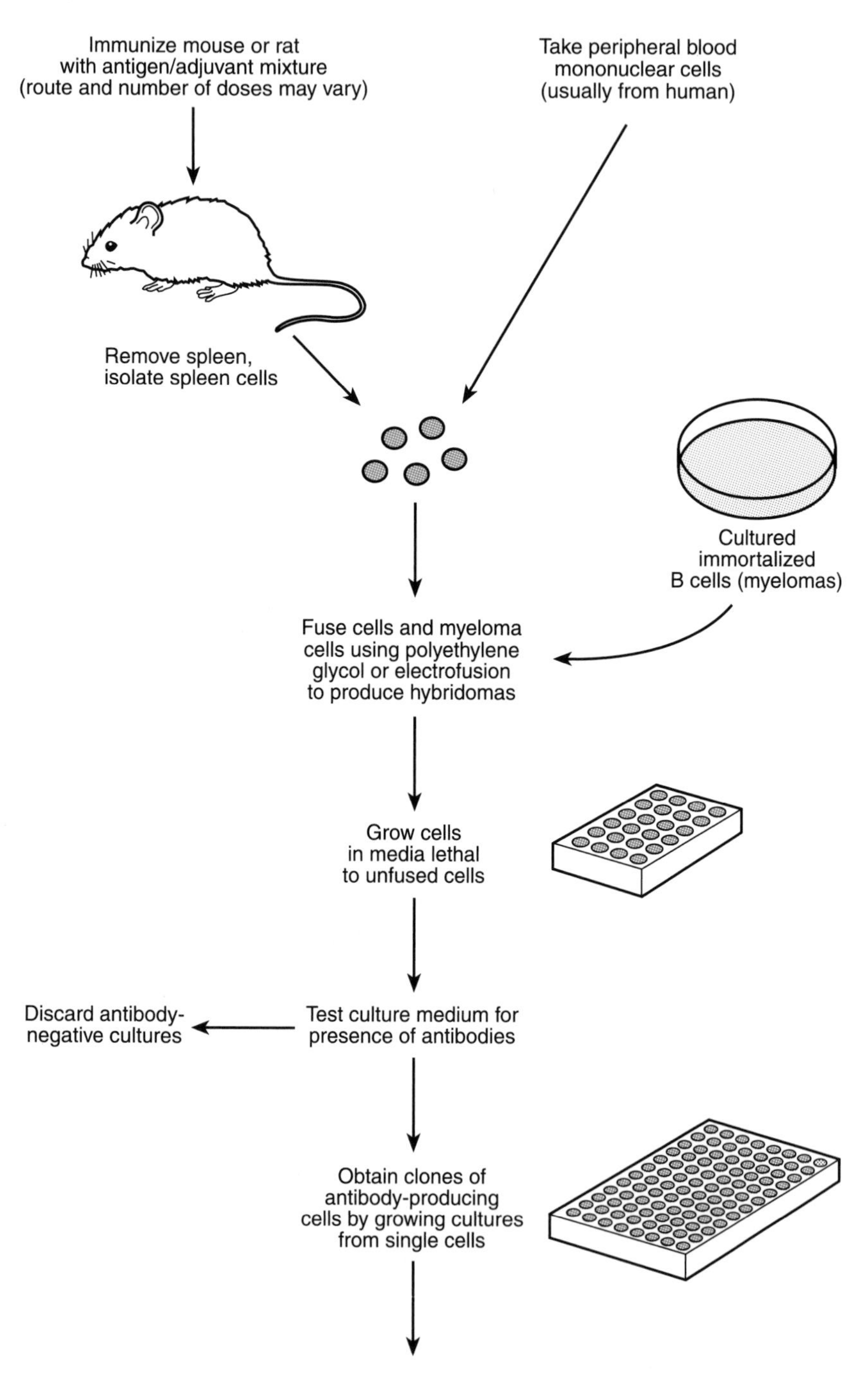

Figure 2.9: Monoclonal antibody production (classical method).

Table 2.7: Production of monoclonal antibodies

Method	Advantages	Disadvantages
Ascites: inoculate hybridomas into mice or rats, harvest 'ascitic' fluid from resultant tumor	'Traditional' method, low cost, high levels of antibody produced	Live animals used, some hybridomas may not form ascites, ascitic fluid may contain many impurities, yield may vary
Culture fluid: harvest from hybridoma cultures	Low cost, simple	Antibody produced at low levels, may require concentration
Bioreactor: various commercially available systems	Continuous production, high levels of antibody possible with some systems	Expensive, may require adaptation of hybridomas

In general, production of ascitic fluid is becoming less common and use of bioreactors is increasing.

monoclonal antibodies to these multiple antigens from the same immunizations. Once hybridomas producing a desired monoclonal antibody are available, useful quantities of antibody can be prepared by any of a number of approaches, summarized in *Table 2.7*. More recent work has made possible the direct production of monoclonal antibodies from circulating B cells, fused to produce hybridomas as above. This has made possible the direct production of monoclonal antibodies from human sources.

Alternatively, immunoglobulin mRNA purified from mouse spleen cells or hybridomas can be expressed in bacteria using recombinant vectors based on filamentous bacteriophages, particularly M13 or fd (*Figure 2.10*). This latter system is now available commercially as the 'recombinant phage antibody system' (RPAS, Pharmacia Biotech), and avoids the use of live animals or eukaryotic cell culture. Using this method, it is possible to produce very large libraries of antibody-producing bacteriophages. These may be derived from immunologically naïve sources (reflecting the general range of antibodies present), or may be taken after immunization or disease. In the latter case, the resultant 'biased' library will be enriched for antibodies binding to the agent to which an immune response was present. It is also possible to mutate or recombine the antibody genes and express them in the phage to allow selection of antibodies with increased affinity, mimicking the affinity maturation that occurs during the natural immune response (see Section 2.3). Unsurprisingly, given its flexibility, use of recombinant phage systems is increasing.

Since its development by Köhler and Milstein in the early 1970s, monoclonal antibody technology has had a revolutionary impact on many areas of science, not least on virology. The number of applications continues to increase.

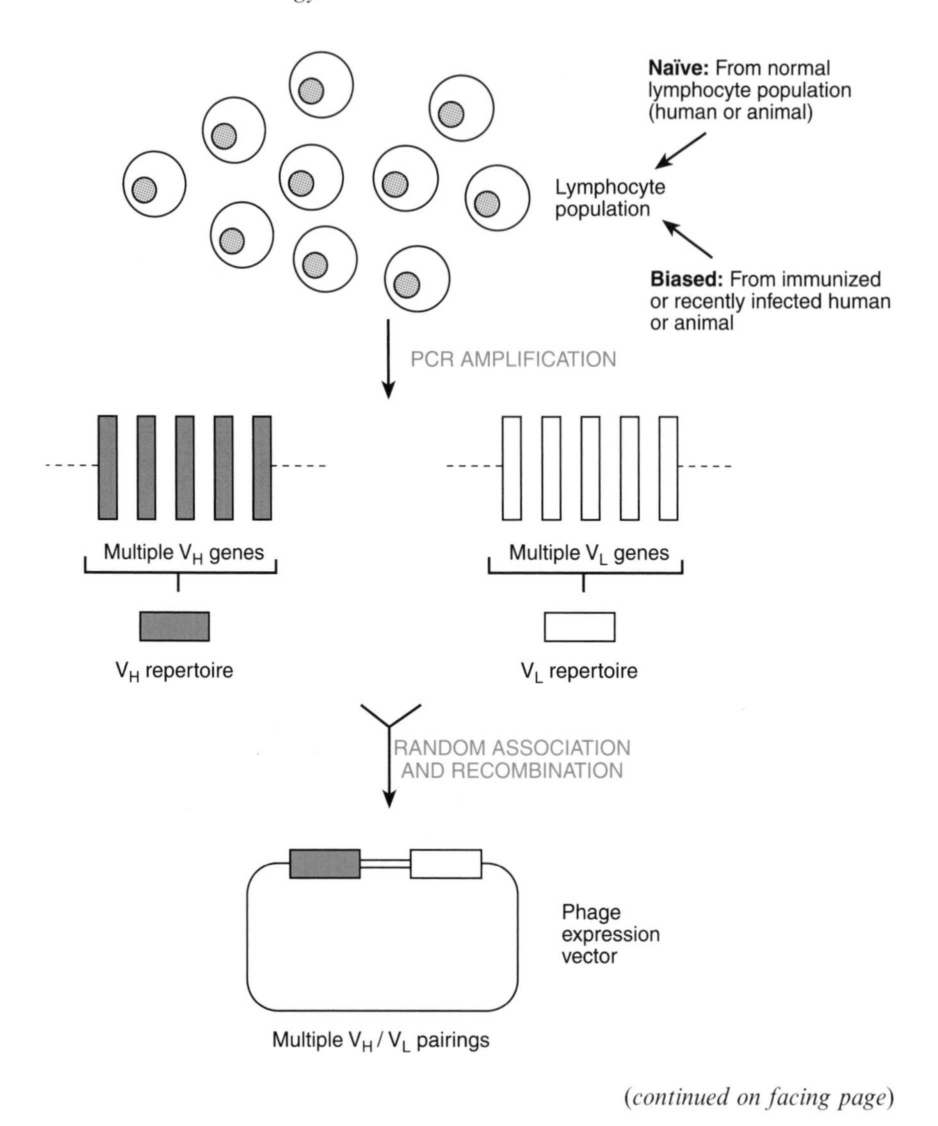

(*continued on facing page*)

Figure 2.10: Recombinant phage antibody production. Derived from Wawrzynczak (1995) *Antibody Therapy*, BIOS Scientific Publishers.

2.9 Passive immunity

The prophylactic/therapeutic use of human immunoglobulins has a long history. They provided some of the earliest protective antiviral treatments, and are still in use for post-exposure prophylaxis with some viruses, including such diseases as Ebola where no other treatment is available. Even where disease is not prevented, the symptoms may be moderated. However, obtaining sufficient supplies of immune serum has always been a problem (as well as extremely expensive), and there is now a trend away

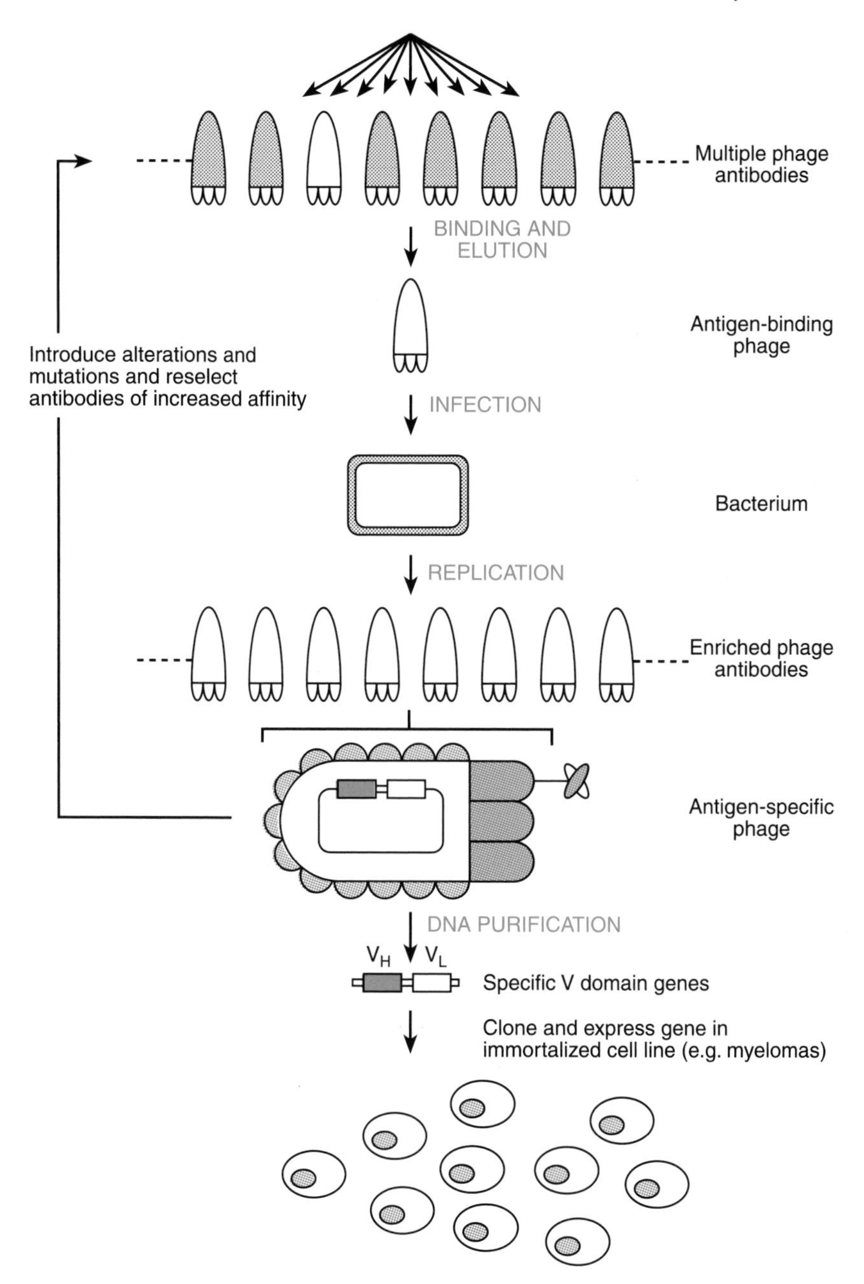

Figure 2.10: *Continued.*

from using human-derived material, due both to the problems of obtaining sufficient material and to the possibility of undetected (possibly even unknown) pathogenic viruses being present. Monoclonal antibodies, which have proven to be of almost unlimited value in viral diagnosis, have yet another application here. They can be produced in effectively unlimited quantities and are far less likely to have pathogenic contaminants. Until recently, most monoclonal antibodies were produced in animals, since the techniques normally used were lethal to the animal producing the original immune cells. Monoclonal antibodies produced in animals are unsuitable for use in humans since they are recognized as 'foreign' and often induce an immune response against the antibody itself. This can ablate protection, and may also cause immunopathological effects. To counter this problem, monoclonal antibodies of human origin can be produced from peripheral blood lymphocytes, and these are appearing in increasingly large numbers. Several groups have attempted to produce variant forms of animal antibodies which could be used in humans by cloning the CDRs of animal antibodies into antibodies of human origin with similar structure, and clinical trials were undertaken with the resulting 'humanized' antibodies. However, with the production of larger numbers of monoclonal antibodies of human origin, humanizing antibodies is proving less popular. It is clear that monoclonal antibodies (of whatever origin) are likely to form the basis of an expanded role for passive immunization in future medicine.

While monoclonal antibodies alone target elements of the immune system, it is possible to enhance the effects of passive immunity by conjugating biologically active substances to a specific antibody. Examples of such molecules include toxins, radioactive compounds, cytokines or antiviral drugs. Toxic agents can be used because effective concentrations are achieved only at the site of antibody binding. For example, bacterial toxins or ricin (a lectin from the castor bean) have been considered for such applications: the natural cell-binding ability of these toxins is blocked or deleted, and they are conjugated to antibody to provide binding only to the target cells. The toxin–antibody conjugates are then taken up by the cell, killing it. Interested readers are referred to another book in this series, *Antibody Therapy*, which covers these matters in greater detail.

Further reading

Banks, T.A and Rouse, B.T. (1992) Herpesviruses – immune escape artists. *Clin. Infect. Dis.*, **14,** 933–941.

Janeway, C.A. and Travers, P. (1997) *Immunobiology: The Immune System in Health and Disease.* Churchill Livingstone/Current Biology Limited/Garland Publishing, London.

Kaplan, M.M. and Webster, R.G. (1977) The epidemiology of influenza. *Sci. Am.*, **237,** 88–106.

Kotwal, G.J. (1996) The great escape: immune evasion by pathogens. *The Immunologist*, **4**, 157–164.

Life, death and the immune system (1993) *Sci. Am.*, **269,** part 3.

Smith, G.L. (1994) Virus strategies for evasion of the host response to infection. *Trends Microbiol.*, **2**, 81–88.

Uren, A.G. and Vaux, D.L. (1996) Molecular and clinical aspects of apoptosis. *Pharmacol. Ther.*, **72**, 37–50.

Winter, G. and Milstein, C. (1991) Man-made antibodies. *Nature*, **349,** 293–299.

Wawrzynczak, E.J. (1995) *Antibody Therapy*. BIOS Scientific Publishers, Oxford.

Electronic resources

For specific sources, use of a search engine such as Alta Vista or Webcrawler will provide links to relevant sites and, since URL addresses change frequently, may be more up to date than the links provided below. Some useful examples are listed below, and provide a variety of links to related material:

Bionet Immunology newsgroup
news: bionet.immunology

Medweb: Immunology, many links
http://www.gen.emory.edu/medweb/medweb.immunology.html

Chapter 3

Vaccines and immunotherapy

Vaccines represent one of the great success stories of virology, with one major cause of human disease eliminated (smallpox, eliminated in 1977), and at least two more targeted for elimination in the near future (polio and measles). The first use of vaccines occurred in the late eighteenth century, and 1997 marked the two hundredth anniversary of Edward Jenner's definitive experiments, using cowpox to prevent smallpox, which gave the world the word 'vaccination'. It should be noted that despite the undoubted successes of vaccination, many important human viral diseases remain for which no vaccine is available. These range from rotavirus diarrhea (*Reoviridae*), a major killer in many countries, for which multiple vaccines are in trial, to the human immunodeficiency virus (*Retroviridae*) and hepatitis C (*Flaviviridae*), for which there is no current candidate vaccine which appears likely to be effective in preventing disease.

Most vaccines are of 'traditional' formulation, and the influence of molecular virology on the design of vaccines has been limited to date. However, a huge range of rationally designed vaccines are now at various stages of clinical testing, with very many more under development. The additional capabilities of such vaccines are discussed below.

3.1 Current vaccines

There are four basic formulations used in currently available viral vaccines, summarized in *Figure 3.1*. These are attenuated live viruses, whole inactivated (killed) viruses, purified subunits (proteins, frequently glycoproteins) of the virus and purified proteins produced from cloned viral genes. Vaccines based on all of these approaches are currently available for human use, and are listed in *Table 3.1*. The newer strategies of expressing viral genes in live vectors or in DNA expression vectors have not yet produced a vaccine available for general use in humans, but are the subject of a great deal of work and are likely to be of great importance in the near future. These are described in detail in Section 3.3.

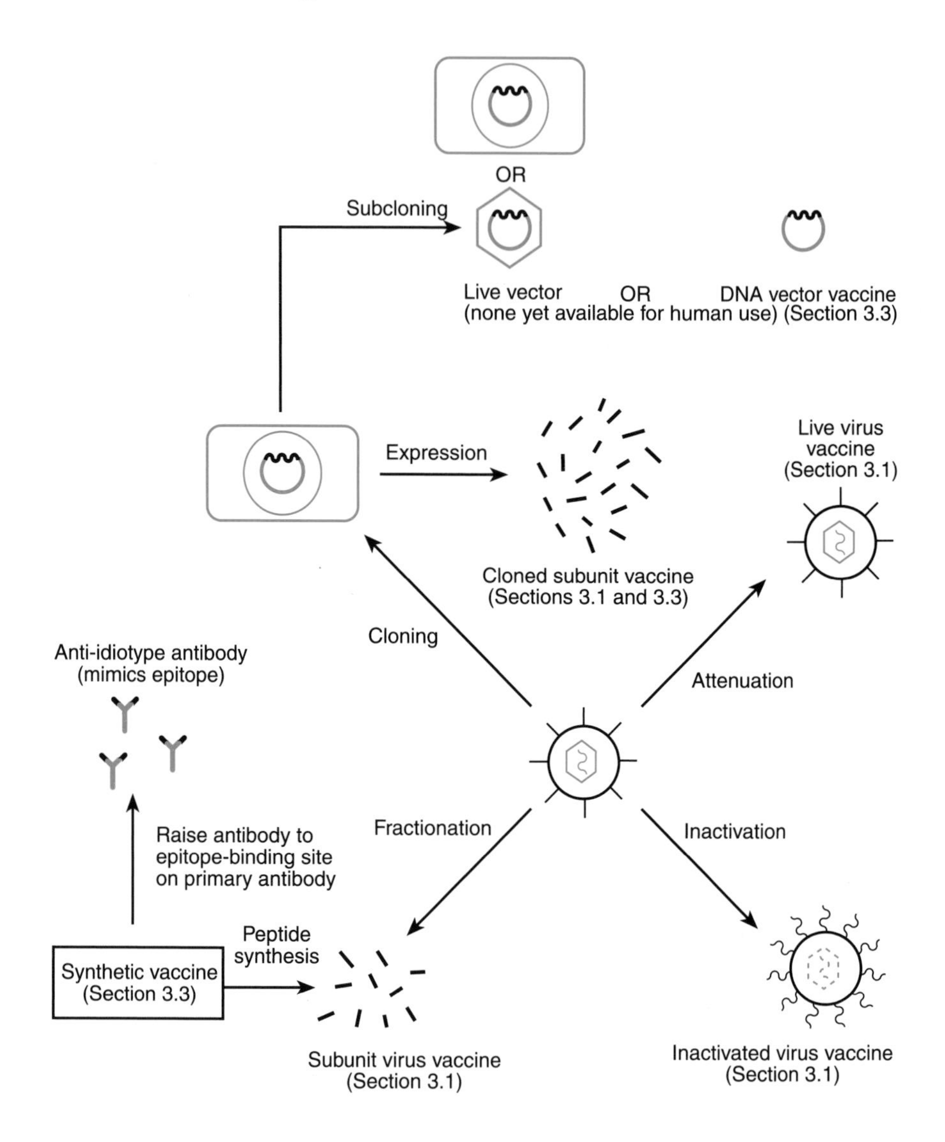

Figure 3.1: Different approaches to vaccine production.

With traditional vaccines based on live viruses, the ability to cause disease is weakened or attenuated by mutation. Historically, such vaccines have owed little to molecular understanding of virology. The mutations are induced by passaging the virus in semipermissive or atypical animal cells (or animals) and/or under altered conditions (e.g. at lower temperatures), selecting variants of the virus which were altered from the optimum for replication in the original host. Nevertheless, they are still able to replicate (if at a reduced level), and are still immunogenic.

Table 3.1 Human viral vaccines

Live attenuated	Inactivated virus	Purified subunit	Cloned subunit
Measles	Hepatitis A	Hepatitis B[a]	Hepatitis B
Mumps	Influenza	Influenza	
Rubella	Japanese encephalitis		
Polio	Polio		
Smallpox[a]	Rabies		
Varicella	Tick-borne encephalitis		
Yellow fever			

[a]No longer in general use.

The oral polio vaccine (OPV) provides an excellent illustration of the development and use of a 'traditional' live attenuated vaccine. It was developed in the 1950s and is a mixture of three serotypes of human poliovirus, attenuated by passage in monkey cells. It is well tolerated, gives long-lasting immunity and is highly protective. It has been used world-wide since 1963 with major benefits for public health (see *Figure 3.2*). However, all live virus vaccines are subject to a range of concerns (see *Table 3.2*), and these are well illustrated by OPV. After administration, the viruses present in OPV can revert to virulent form and can then be transmitted by the vaccinee, causing disease. Sequencing of the genomes

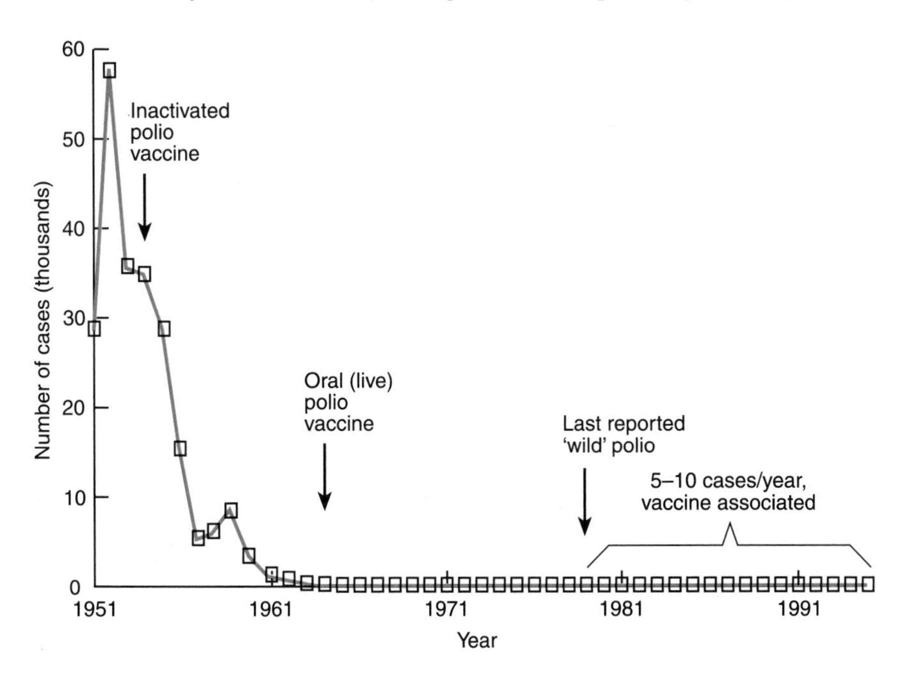

Figure 3.2: Reported cases of poliomyelitis in USA.

Table 3.2: Advantages and disadvantages of vaccine types in current use

Vaccine type	Advantages	Disadvantages
LIVE	Relatively simple development and production Low cost possible Stimulates cell-mediated immunity Can be highly immunogenic and protective, giving long-lasting protective immunity (mimics natural disease) May require fewer inoculations than other systems	May cause mild form of disease Reversion to virulence Virus shedding Requirement for continuous refrigeration ('cold chain') Possibility of contaminating viruses Contains viral genome, which may be pathogenic or oncogenic in some systems
INACTIVATED	Relatively simple development and production Quite low cost No reversion or spread of virus Stable, may not require cold chain Low risk of live virus contamination	Requires culture system able to produce virus in large quantities May require adjuvants May not be highly immunogenic May require frequent revaccination Contains viral genome, which may be pathogenic or oncogenic in some systems
PURIFIED SUBUNIT	Can target immune response to part of the virus No reversion or spread of virus Stable, may not require cold chain Very low risk of live virus contamination	Quite complex to produce May be expensive Require culture systems able to produce virus in large quantities Require adjuvants May not be highly immunogenic May require frequent revaccination May contain whole virus or viral genome
CLONED	Quite low production costs are possible Relatively simple scale-up of production Possible to optimize or combine antigens No reversion or spread of virus Stable, may not require cold chain No risk of live virus contamination	Expensive and complex development Require adjuvants May not be highly immunogenic May require frequent revaccination

has shown that the type three vaccine strain, the strain most commonly associated with reversion to virulence, has less changes from the wild-type parent than do the type one and two vaccine strains. Biotechnology offers a number of routes to improving even a pre-existing live vaccine. Knowledge of the genome sequences of the viruses present in the OPV could allow mutations associated with non-reversion of the type one and two strains to be introduced into the type three genome – the 'defined attenuation' approach, as against the 'random attenuation' of the original vaccine. An alternative approach is to clone type three antigen genes on to a core of a 'less problematical' poliovirus strain, representing a limited form of the live vector approach, discussed in more detail in Section 3.3. In live vaccines in general, the mutations causing attenuation are usually alterations of single bases (point mutations), which can revert easily since only a single base change is required. Deletion or addition of groups of bases, while rarer, may occur, and such a mutation is less likely to revert. It is now possible to introduce specific deletions of such groups of bases, making reversion far less likely. With poliovirus the viral genome is RNA and, as a result, it has a relatively high mutation rate (see Section 1.4.5). The possibility of mutations in the seed stocks used to prepare the virus has always been a concern but has now been greatly reduced by cloning the whole genome of the seed strains and maintaining them as DNA copies (complementary DNA, or 'cDNA' molecules). If these copies are introduced into cells, the virus can be produced. This procedure is referred to as 'genetic stabilization'.

Since live vaccines are grown in animal or human cells, contaminating or endogenous (literally, 'produced from within') viruses can be a problem. This was observed with early batches of OPV, where simian virus 40 (SV40), a member of the *Papovaviridae* and a highly transforming virus, was also produced by the monkey cells used to grow the virus. SV40 was also present in some early batches of the inactivated (killed) polio vaccine, since SV40 was not inactivated by the treatments used to inactivate the poliovirus present. Long-term monitoring has shown no SV40-associated disease in recipients of contaminated vaccine, but the episode indicated the need for great care in the use of animal cells for vaccine preparation, and cell types used are now very closely monitored by the relevant authorities.

The concerns surrounding the use of live, classically attenuated, viral vaccines must be balanced against the highly immunogenic nature of live viral vaccines and the major health benefits that they have brought. Again, using the example of poliovirus, while inactivated vaccines have been used, the most commonly used formulation has been the (more immunogenic) trivalent live vaccine. Vaccination against polio has eliminated the disease in the Western hemisphere and, in 1996, only 3853 cases of polio were reported world-wide. This is a success by any standards (see *Figure 3.2*), and a target for the world-wide elimination of polio has been set by the World Health Organization. It must be borne in mind that the relatively crude attenuated live vaccines now available have saved many lives. This has yet to be achieved by any advanced biotechnological vaccine.

Another approach to achieving attenuation which is under study for some viruses (notably rotavirus) is to substitute genes controlling virulence with those from related animal viruses. It is known that such changes can alter virulence, not least because such reassortment is thought to be how influenza causes world-wide pandemics.

By understanding the molecular factors underlying virulence, it should be possible to make defined alterations in specific genes which will attenuate a virus. Examples could include the genes which control latency in herpesviruses or neurovirulence in poliovirus.

Since many workers believe that live attenuated vaccines can provide a broader and more protective immunity than do subunit vaccines, intensive studies are under way to identify and modify virulence determinants in a number of viruses, including HIV. Defined attenuations are likely to form the basis of future live vaccines, and are also important in the development of viral vector systems, discussed in Section 3.3.

As with live vaccines, whole inactivated ('killed') vaccines produced by culturing and then chemically inactivating (killing) the virus have owed little to molecular virology, but have been very successful in controlling many viral diseases. While the use of whole inactivated virus is usual, not all viral proteins are immunogenic and the presence of other viral proteins may dampen the immune response to the protective antigens. As a result, some vaccines have been produced by purification of immunogenic viral components, including those for influenza and hepatitis B. In the latter case, hepatitis B surface antigen was originally purified from the blood of carriers of the virus, but this has now been superseded by a cloned antigen produced in yeast cells, avoiding any problems of residual infectivity. This is the only 'cloned subunit' vaccine currently available for use in humans. However, in those vaccines currently under development, the component immunogens are more likely to be derived from cloned viral genes expressed in heterologous cells than from the virus itself. Such an approach simplifies the production of these immunogens on an industrial scale, and also allows manipulation and optimization of protein immunogenicity. For those viruses which do not grow or grow very poorly in tissue culture, such as papillomaviruses and EBV, cloned viral genes may offer the only viable approach. The advantages and disadvantages of the types of vaccine available are summarized in *Table 3.2*, and the basic principles of cloning and expression of viral genes are covered in Section 5.3. Proteins expressed in prokaryotic or eukaryotic expression systems may be used as antigens, but prokaryotic systems, while relatively straightforward to use, produce proteins which may differ from the 'authentic' protein. Eukaryotic systems are more complex but are capable of producing proteins more like those of the virus itself, which are often more immunogenic (see Sections 5.2 and 5.3). As a result, techniques to scale up eukaryotic systems for industrial vaccine production are under active development.

3.2 Adjuvants

Vaccines consisting of purified protein (rather than whole inactivated viruses) frequently require the presence of an adjuvant to induce a protective immune response, since proteins in solution may be poorly immunogenic. The functions of an adjuvant can be divided as follows.

- Presentation of antigen in a particulate form, enhancing uptake by antigen-presenting cells.
- Direct stimulation of the immune response, often mediated by cytokines or antigen-presenting cells.
- Localization of antigen to the site of inoculation, also known as the 'depot effect', providing an intense local immune stimulation rather than a diffuse generalized stimulus.
- Targeting of antigens to particular pathways.

The relative importance of each of these effects varies with different adjuvants. Of all those adjuvants which are available (a selection of which are shown in *Table 3.3*), alum (microparticulate aluminum hydroxide gel) is the only clinically significant adjuvant currently licensed for use in humans. Antigenic proteins are adsorbed on to the surface of the alum particles, and the depot effect is thought to be important for alum-mediated immunogenesis. However, alum does not appear to modulate the immune response directly.

Immunization of animals for the purpose of raising antibodies has traditionally used Freund's adjuvant. The complete form contains killed mycobacteria in a water-in-oil emulsion. While the water droplets of the emulsion serve to give a particulate character, the mycobacterial component is a potent immunomodulator – rather too potent, since it is the cause of significant toxic effects, making it unacceptable for human use. Freund's adjuvant was developed when understanding of molecular events was very limited, and the nature of the immunomodulatory components of this and other crude adjuvants was poorly understood. More recent studies identified muramyl dipeptide and lipid A as bacterial components mediating an immune response in crude adjuvants. These were still too toxic for use in humans, but less toxic derivatives have been produced which show reduced toxicity while retaining immunomodulatory effects.

Many experimental adjuvants have been evaluated in animal systems and in humans. However, none are yet available for general use. As with vaccines themselves, many promising developments are based on our increasing understanding of events at the molecular level. Again, as with vaccines, none of these have as yet proved widely useful despite promising results in some clinical studies. There are many examples of 'second generation' adjuvants, and some which are available are listed in *Table 3.3*. The basic functions of typical components of these adjuvants are shown in *Table 3.4*. One clear advantage of some second generation adjuvants is their ability to stimulate a $CD8^+$ T-cell response. These cells are stimulated by antigen presented in combination with proteins of the MHC-I system expressed by all cells (see Section 2.2.1) and, until relatively recently, it was one of the 'facts' of vaccine design that proteins

Table 3.3: Examples of adjuvants for antibody production or immunization

Product name	Chemical composition	Interaction with antigen
Freund's adjuvant[a]	Complete: mineral oil + killed mycobacteria Incomplete: mineral oil only	Water-in-oil emulsion
TiterMax™	Non-ionic block co-polymer bonded to silica microparticles + squalene + emulsifiers	Hydrophilic binding; water-in-oil emulsion stabilized by microparticles
Ribi adjuvant	Squalene + Tween 80 + monophosphoryl lipid A + trehalose dimycolate	Oil-in-water emulsion
Alum	Aluminum potassium phosphate microcrystals in suspension	Adsorption
Aluminum hydroxide; Alhydrogel™	Colloidal gel suspension of microcrystals used alone or mixed with muramyl dipeptide	Adsorption
SAF-1	Squalane/squalene + threonyl-muramyl dipeptide	Oil-in-water emulsion
Adjuvax™	Micron-sized branched β-glucan particles	Adsorption in aqueous colloidal suspension
Opti-Vant™	β-glucan particles	Adsorption in aqueous colloidal suspension
Quil A	Purified saponin	Surfactant–antigen complex
DDA	Dimethyldioctadecyl ammonium bromide (lipoidal quaternary ammonium compound)	Ionic and hydrophobic interactions
MF-59	Squalene + Tween 80 + Span 85	Oil-in-water emulsion
Provax	Squalane + Tween 80 + non-ionic block co-polymer	Oil-in-water emulsion with block co-polymer-mediated hydrophilic binding

Modified from Harper (ed.) (1993) *Virology Labfax*, pp. 192–193, BIOS Scientific Publishers.

presented via the MHC-I/CD8$^+$ pathway had to be synthesized inside the presenting cells (see Section 2.2). This was perceived as an important problem for any subunit (or inactivated) vaccine directed against a disease where cell-mediated immunity was important, which includes many virus diseases, since it was thought that these antigens would only be presented to CD4$^+$ T cells via the MHC-II pathway (see Section 2.2.2) and would not induce a significant cytotoxic T-cell response. There is now convincing

Table 3.4: Functions of adjuvant components

Component	Examples
Particles	Alum, ISCOMS[a], emulsions, liposomes/proteosomes, block co-polymers[b]
Stabilizers[c]	Detergents, surfactants, block co-polymers[b]
Immunomodulators	Muramyl di- and tripeptide derivatives, lipid A derivatives, block co-polymers[b], saponins[d], cytokines

[a]ISCOMS, immune stimulating complexes. Virus-sized cage-like structures formed of saponins and lipid.
[b]Block co-polymers, linear polymers of clustered hydrophilic and hydrophobic monomers, with central hydrophobic regions flanked by smaller hydrophilic termini.
[c]Stabilizers, may be required to stabilize emulsions or hydrophobic structures in an aqueous environment.
[d]Saponins, complex glycosides, purified from vegetable resins. Early crude preparations contained both adjuvant and toxic saponins, but non-toxic saponins with adjuvant effects have been identified, most notably QS-21, opening the way for human use.

evidence that some adjuvants (*Table 3.5*) can allow proteins to enter the MHC-I-mediated pathway, apparently by entering antigenic proteins directly into the cytoplasm. This makes the development of effective subunit vaccines far more likely in many cases. There are also reports that unconjugated peptides may combine with MHC-I molecules and induce a cytotoxic T-cell response if the epitope represented is selected correctly. It is clear that both antigen and adjuvant require careful selection.

The importance of individual components of the immune response varies between different virus infections; resolution of infection by viruses that usually remain associated with infected cells is likely to require a strong CMI, while antibody is more important in resolving infection by viruses that produce high levels of free virus particles. Production of protective immune responses to these different infections requires 'tailoring' of vaccine immunogenicity (see Section 3.5) using different

Table 3.5: Adjuvants known to induce an MHC-I-mediated ($CD8^+$) T-cell response

ISCOMS[a]
Saponin QS-21[a]
Squalene[b]/MTP-PE[c]/surfactant
Liposomes/proteosomes
(VLP–Ty[d], keyhole limpet hemocyanin[e])

[a]ISCOMS, saponins, see *Table 3.4*.
[b]Squalene, a hydrophobic unsaturated shark liver oil.
[c]MTP-PE, muramyl tripeptide-phosphatidylethanolamine, a non-toxic lipophilic muramyl dipeptide derivative with large hydrophobic groups which serve to anchor it to hydrophobic particles in an emulsion.
[d]VLP-Ty, not an adjuvant, but rather a fusion vector (see Section 3.3) with adjuvant effects. TyA protein produced in yeast cells containing a truncated form of the Ty genetic element is modified to produce foreign protein fused to TyA. These fusion proteins form particles of similar size to viruses, containing the foreign protein (virus-like particles, or VLPs).
[e]Keyhole limpet hemocyanin, a highly immunogenic, very large protein with known immunostimulatory properties, to which peptide antigens are chemically linked.

combinations of antigen and adjuvant, and this will be essential to the production of effective new vaccines.

3.3 Approaches to vaccine development

When considering the current efforts to develop antiviral vaccines, it is important to make a distinction between research aimed at developing a vaccine to protect against infection or against disease, and the (very interesting but distinctly different) research aimed at increasing understanding of the effects and mechanisms of protective immunity. Much of the work that is taking place, even within industrial laboratories, falls into the second category, and while such work may provide the basis for future generations of vaccines, direct clinical applications are still remote.

Approaches to vaccine production based on the application of biotechnology are shown in *Table 3.6*, and are discussed in more detail below.

The use of purified antigenic proteins produced from cloned viral genes to prepare subunit vaccines has been discussed in Chapter 2, and cloned antigens have been used in many clinical trials as well as in the only biotechnology-derived vaccine currently available for human use, that for hepatitis B. A variation on this theme is to use peptides, either synthesized from cloned genes or produced chemically, which correspond to known epitopes of viral antigens (see Section 2.7). This approach may be particularly promising for the induction of cell-mediated immunity, since the natural presentation of proteins to T cells is in the form of peptide fragments. Peptides are less suitable for the induction of antibodies: despite the ability to select peptides corresponding to highly immunogenic linear epitopes, they have a very limited ability to present more complex epitopes (see Section 2.7), and lack any post-translational processing, such as glycosylation. Following early unpromising results with peptides, they are now more commonly formulated using a fusion vector approach. In such a system, the gene coding for the peptide is linked directly to a gene for a carrier protein. The protein synthesized by this hybrid gene contains the peptide as part of a modified carrier molecule. A less commonly used alternative approach is to fuse a synthetic peptide directly to the carrier. Both approaches have been used in clinical trials. Where a fusion vector is

Table 3.6: Biotechnological approaches to vaccine development

Defined attenuations (specific genetic alterations) of live viruses, including single-cycle infections (DISC)
Purified antigenic proteins (whole or fragments) produced from cloned viral genes (subunit vaccines)
Antigenic peptides or protein fragments attached to carrier proteins (fusion vectors)
Antibody molecules that mimic viral antigens (anti-idiotypic antibodies)
Genes from one virus expressed in another or in bacteria (live vectors)
Virus genes expressed in a vector incapable of complete replication (gene vectors)
Purified plasmid DNA expressing viral gene under eukaryotic promoter control (DNA vector)

used, the carrier is often selected for its adjuvant effect (see Section 3.2). Examples include a yeast system, where the fusion protein assembles into 70-nm 'virus-like particles' or the direct fusion of smaller synthetic peptides to keyhole limpet hemocyanin. Both may be immunogenic for cell-mediated and serological immunity. In fact these are two of the systems that appear to be able to enter peptides into the MHC-I/CD8$^+$ (cytotoxic T cell) pathway, presumably by entering protein into the cytoplasm, as discussed in Section 2.2.1. One potential advantage of fusion vectors is that multiple peptides can be inserted into one fusion vector, allowing multiple epitopes to be presented. These can be epitopes directed towards different arms of the immune system (cell-mediated and humoral) or epitopes from different proteins or even different organisms. This multivalent approach is the subject of a great deal of interest, since it could (in theory) allow one vaccine to protect against a range of diseases or against different serotypes of the same virus. Of course this approach is not entirely novel, since mixed vaccines are already in use, for example in the trivalent oral polio vaccine (see Section 3.1) and in combined vaccines such as the currently available measles/mumps/rubella (MMR) formulation. One area that has attracted particular interest is the inclusion of multiple variants of the HIV gp120 V3 loop (see Section 2.7) in a single vaccine to counter the highly variable nature of HIV. However, peptides are relatively small, and even minor, localized changes in the corresponding viral protein could result in vaccine failure.

A variation on peptide vaccines is the use of antibody molecules raised against antigen-binding sites of antiviral antibodies (anti-idiotypic antibodies). These mimic the conformation of the viral antigen, but this method does not appear to hold much immediate promise for viral vaccination and may be more applicable where an antigen cannot be produced by cloning, for example the polysaccharide epitopes which distinguish melanoma (skin cancer) cells.

All of the above approaches rely on the presentation of non-replicating proteins to the immune system. An alternative approach is to express the gene coding for the antigen in a virus or bacterium that can make the antigen inside the vaccinee. While bacterial vectors (particularly those based on attenuated strains of *Salmonella*) appear promising, given the nature of this book we will concentrate on the range of live viral vectors currently under evaluation.

The vector virus is attenuated, either in the classical way (see Section 3.1) or by the deletion or site-directed mutation of viral genes. The inserted gene is expressed in the vaccinee's cells. The protein produced in this way acts as the immunogen, stimulating both cell-mediated and humoral immunity. Much early work used the poxvirus vaccinia, which was used as the vaccine for smallpox. However, the strain used as a vaccine vector was frequently a virulent strain rather than the accepted vaccine strain. An altered vaccinia virus (NYVAC) has been constructed for use as a vaccine vector, which has 18 open reading frames (genes) deleted from six different regions of the genome. Despite this, it is still capable of growth in some cell culture systems and is immunogenic when used as a vector. The related canarypox virus is non-pathogenic for humans even in its unaltered form, and a 'doubly safe' version of this virus has been produced (ALVAC) for use as a live vector. ALVAC is made doubly safe both by deletion of multiple open reading frames and by its inherent non-pathogenicity for humans. Such developments could make poxvirus vectors more acceptable for human use, although concerns remain about any poxvirus infection in an immunocompromised patient.

An alternative vector is the attenuated (live) Oka vaccine strain of varicella-zoster virus (VZV, human herpesvirus 3). Multiple genes have been expressed in this system, but its usefulness is limited by several factors. It is a classically attenuated virus, and the basis of its attenuation is poorly understood. In addition, it is known to establish latent infection following vaccination. Finally, although now licensed for general use in the United States, many countries world-wide have not yet approved even the unmodified virus.

When selecting a virus for use as a live vector, a large genome (like that of pox or herpesviruses) allows the insertion of multiple genes, again raising the possibility of multivalent vaccines. Even with smaller viruses it is possible to insert fragments of genes coding for a series of immunogenic peptides, known as the 'string of beads' approach. Relative sizes of viruses under study for use as live vectors are shown in *Figure 3.3*. While herpesviruses have large genomes, the possible effects of live vectors establishing latent infection (see Section 1.6) require investigation. Similar problems attend the use of retroviral vectors, compounded by the possible oncogenicity of the vector. Both polio and adenovirus vectors could allow oral vaccination (see Section 3.6), but the small size of the poliovirus genome could be problematic and would require the expression of peptide rather than protein antigens. The even smaller adeno-associated parvovirus vector system would also have this problem, as well as any associated with its integration into the cellular genome. Adenoviruses, with relatively large genomes and mild pathogenicity even in their unaltered form, are proving an attractive system for the development of live vectors for use *in vivo*, particularly since adenovirus vaccines are already in use by the US military. However, while adenovirus vaccines seem to be safe, there are concerns over inflammatory responses and possible oncogenicity (see Section 1.7).

A variation on the use of live virus vectors is the use of replication-defective variants grown in cell lines expressing viral genes or otherwise providing the 'missing' functions for viral replication. The viruses produced in such systems may infect other cells but are not then able to produce infectious progeny virus. Such systems are referred to as 'gene vectors' since they are not fully 'live' and can only undergo very limited replication in the vaccine recipient. In fact, some of the 'live' vectors now under study (such as ALVAC) appear to fall into this category. However,

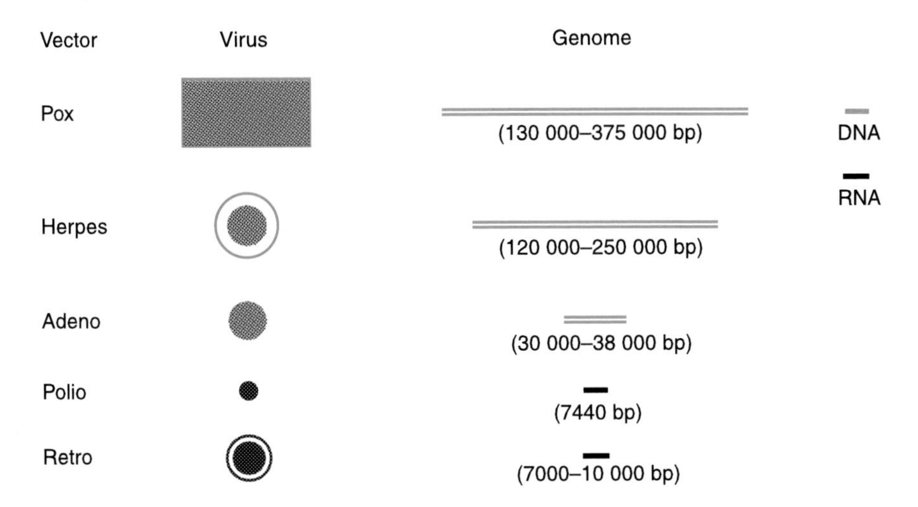

Figure 3.3: Viruses used as live or gene vectors.

since proteins are produced within the cells of the vaccinee, both live and gene vectors can stimulate the full range of cell-mediated immunity, as well as humoral and non-specific immune responses. As well as potential uses as vectors, replication-defective viruses may be used to stimulate immunity to the parental virus. This is represented by the DISC (disabled infectious single cycle) vaccine approach. In this, an essential gene is deleted and the virus grown in cells expressing the missing gene. The resulting, fully infectious virus is used as a vaccine. It can infect cells within the body and replicate, thus calling in all elements of the immune response; however, the virus produced is not infectious since the essential gene is not present. A DISC vaccine for genital herpes simplex infection (HSV-2) lacking the glycoprotein H (gH) gene is now in clinical trials.

3.3.1 Nucleic acid vaccination

Following reports of an immune response resulting from injection of plasmids expressing genes, work in 1993 showed that expression of the influenza nucleoprotein (NP) gene could protect mice from infection. Since this time, work in this area has expanded dramatically due to the simplicity and wide applicability of the technique. The basic technique involves the construction of a plasmid coding for an antigen under the control of a strong promoter (usually the CMV immediate–early promoter) (see Section 5.2). This is then introduced, usually into muscle tissue, and can induce the full range of immune responses. Alternative delivery systems are under investigation, particularly delivery on small gold beads (fired from a helium-powered 'gene gun') and direct application of the DNA to mucosal surfaces.

The initial finding was surprising since nucleic acids are readily degraded within the body, and it was initially estimated that only one in a million plasmids enter cells and express the viral gene. The use of DNA vaccines has been referred to as a 'revolution' in vaccinology, and has made possible a far more rapid progress from genetic information to candidate vaccines than was possible before. A very wide range of DNA vaccines are under development, and clinical trials for HIV have been in progress since 1995. Despite some concern that the system might be less applicable outside the mouse model where early evaluations were performed, there is now evidence of protection in studies with pigs, and the results of the human trials are awaited. It should be noted that such studies once more confirm the need to avoid DNA contamination of protein-based vaccines.

One problem (avoided by DNA vaccines) with any approach using fusion, live, or possibly even gene, vectors is the presence in the vaccine of the immunogens of the vector itself. Some of the live vectors under study are viruses to which a significant section of the population have been exposed. A strong immune response to the vector could occur, boosted by pre-existing immunity to these antigens, and this could result in damping of the immune response to the novel (vaccinating) antigens. Even where immunity does not already exist (as would be the case with many fusion vectors), the first such vaccine to be used might reduce the response to the novel components of any future vaccination using the same vector system. While some workers discount this possibility or even suggest a potentiating effect on immunogenicity, the issue has not yet been fully resolved.

There are so many vaccines in trial or under development that it would be almost impossible to list them all. The potential properties of some of the vaccine types in development are shown in *Table 3.7*. However, given the level of interest in developing an HIV vaccine, it is not surprising that almost all of the approaches listed in *Table 3.6* are being evaluated in attempts to obtain a protective immune response against HIV.

3.4 Possible pathogenic effects of biotechnology-derived vaccines

Reversion to virulence by attenuated live vaccine is discussed in Section 3.1. However, subunit vaccines can also have pathogenic effects. A vaccine containing inactivated respiratory syncytial virus (RSV) has been shown to increase the severity of subsequent infection. Formalin treatment during vaccine preparation reduced the immunogenicity of the viral glycoproteins, and the resulting non-neutralizing immunity potentiated RSV disease. Similar effects are known for LCMV in animal models. Stimulation of the 'wrong type' of immunity can also have adverse effects, for example in the potentiation of virus entry across the respiratory epithelium by secretory IgA, an effect seen with EBV. There is also the possibility of direct pathogenic effects: one major concern with subunit vaccines based on the HIV gp120 glycoprotein is that this protein has been

Table 3.7: Potential properties of vaccine types in development

Vaccine type	Adjuvant required?	Cell-mediated immunity?	Multivalency?[a]
Live (defined attentuation)	–	+ + +	–
Subunit protein	+ +	±[b]	+
Peptide	+ + +	±[b]	+ +
Fusion vector	+[c]	±[b]	+ +
Anti-idiotype	±	±[b]	–
Live vector	–	+ + +	+ + +
Gene vector	±	+ +	+ + +
DNA vector	–[d]	+ +	±

[a]In a single system, without mixing of different vaccine preparations.
[b]Depending on adjuvant used.
[c]Vector may have inherent adjuvant activity.
[d]Potentiating drugs may be co-administered, and are thought to function by enhancing DNA uptake.
This table provides only a general guide – individual exceptions exist in most cases.

implicated in the induction of apoptosis in cells of the immune system. This is clearly not a desirable property for a vaccine to exhibit. All this means that it is essential that vaccines are evaluated thoroughly for safety. Indeed, the level of concern in this area is reflected in the fact that clinical trials address the issue of safety before efficacy against disease is even considered. In addition to antigen effects, toxicity associated with experimental adjuvants can be significant, emphasizing again the need for careful testing of novel vaccine formulations.

3.5 Tailoring of the immune response to vaccination

It is clear that different infections are controlled by different elements of the immune system, and that in some cases a humoral antibody response is insufficient. For example, it is clear that reactivation of herpesviruses can cause disease even in the presence of high levels of neutralizing antibody in the blood. In this particular case, the tightly cell-associated nature of the virus is thought to require the cell-mediated cytotoxic elements of the immune system to control the infection. Conversely, viruses which exist free in the blood, such as hepatitis B, may be susceptible to antibody-mediated control. Thus it can be seen that the specific nature of the immune response induced by a vaccine must be appropriate to the pathogen. A great deal of work is currently under way aimed at identifying the 'correlates of immunity' for viral (and other) diseases, with the intention of specifically stimulating either T_H1 or T_H2 responses (see Section 2.2.2) so as to favor the development of either cellular or humoral immunity. Such vaccines will use tailored adjuvants (see Section 3.2) and delivery systems, including cytokines and other immunomodulatory elements. However, it should be noted that the immune system is highly complex, and attempts to provide a universally effective 'tailoring' are likely to prove very difficult.

3.6 Alternative delivery systems

Traditionally, most vaccines have been administered by inoculation. However, experience with the live polio vaccine (for which oral infection is the natural route) has shown the increased ease of administration of a vaccine that can be given by the oral route. Furthermore, oral delivery is effective at stimulating mucosal immunity in the gut, which is thought to be very important in protecting against poliovirus infection. However, live poliovirus naturally infects by this route. For a subunit vaccine, specialized systems must be used, since the gut is optimized to digest proteins rather than mount an immune response to them. A range of approaches to getting protective vaccination by oral ingestion are under development, including (not unreasonably) the use of vectors based on live poliovirus. Such approaches are listed in *Table 3.8*. However, both vaccinia poxvirus and adenovirus may be used to vaccinate by the oral route, although neither causes pathogenic infections by this route. In the former case, a recombinant rabies/vaccinia vaccine is contained in bait capsules which are laid for wild animals, including foxes and raccoons, and it is thought that the virus may enter the blood via cuts in the mouth. In the latter, in a vaccination program restricted to the US military, 'enterically coated' tablets containing freeze-dried virus are used. These release virus into the intestine, giving an asymptomatic disease that induces a protective immune response.

Mucosal immunity is an area under intensive investigation, since it has the potential to stop a virus from even entering the body ('sterile

Table 3.8: Examples of oral vaccine delivery systems

Viral systems
Poliovirus vectors
Adenovirus vectors
(Poxvirus (vaccinia) vector in encapsulated form)
Bacterial systems
Salmonella vectors
Fusion systems
Cholera toxin B subunit vectors
Tetanus toxin C subunit vectors
Passive carrier systems
Enterically coated tablets containing lyophilized virus
Fusogenic liposomes/proteosomes
Amino acid–gum condensates (stable at acid pH in stomach, dissociate in neutral/alkaline intestinal environment)
Dual emulsions[a]
Cochleates[b]

[a]Small hydrophilic droplets containing antigen within larger hydrophobic droplets, stabilized by block co-polymers.
[b]Tightly folded lipid structures.

immunity') and can be stimulated at almost any mucosal surface. Such immunity involves secretory IgA (see Section 2.3) as well as infiltrating antibodies and cells from the blood, and can be stimulated by inoculation of vaccine onto the mucosae. There is good evidence that mucosal immunity may be required for optimal protection against viruses that infect via such a route, including the oral, respiratory and genital routes of infection. However, there is some evidence that the resultant immunity may be restricted to the specific site of inoculation, possibly requiring the use of multiple vaccination sites, some of which are likely to prove highly unpopular if used in a clinical setting. Despite this, studies on optimizing vaccine delivery via almost every mucosal surface are currently under way, often in combination with systemic vaccination.

Another variant delivery system under investigation is 'slow release' systems. These would prevent the need for multiple vaccinations by delivering antigen over a period of weeks from within a biodegradable polymer implanted at the site of the initial vaccination. Many problems remain in this area, such as how to ensure the stability of the antigen prior to release, when it would be warm and perfused by body fluids, and the possibility of severe localized immune reactions.

It is clear that many of the vaccines now under development will be 'tailored' to produce a spectrum of immunity providing maximum protection against a particular route of infection by an individual pathogen. This will involve careful selection and evaluation of antigen, vector system, adjuvant and route of delivery. All of this will need to be based on an understanding of events at the level of molecular virology.

3.7 Therapeutic vaccination

While vaccination traditionally is considered as a method of preventing disease (prophylaxis), it is also possible to use vaccination to moderate the effects of a pathogen which is already present (therapy). There are a number of diseases where the organism responsible remains present at low levels or in an inactive form and causes disease at a later date; examples include herpesvirus infections and papillomavirus-induced cancers of the cervix. In addition, there are many 'slow' infections where symptoms develop a long time after infection, including the development of AIDS after HIV infection. In all of these cases, the use of a vaccine to boost immunity has the potential to help prevent progression to disease. Candidate vaccines for herpes simplex virus (HSV) and a range of papillomaviruses have been shown to reduce recurrent disease in animal models, and candidate vaccines against HSV and HIV have been investigated in clinical trials. The DISC vaccine approach is under investigation for therapeutic vaccination against HSV (see Section 3.3). It is also possible to use 'classical' vaccines in this role, and extensive studies have been undertaken on the use of the live varicella vaccine to boost immunity in the elderly with a view to preventing zoster (which is caused by reactivation of VZV), although definitive trials are still awaited. While there is as yet no routinely used therapeutic vaccination in humans, it is likely that this approach will play a significant role in disease control in the future. Some vaccines (e.g. rabies) may be used after exposure, but this relies on establishing an immune response before infection becomes established (prophylaxis) rather than being a true therapeutic vaccination.

Further reading

Ertl, H.C. and Xiang, Z. (1996) Novel vaccine approaches. *J. Immunol.*, **156**, 3579–3582.

Langermann, S. (1996) New approaches to mucosal immunization. *Semin. Gastrointestinal Dis.*, **7**, 12–18.

Mackett, M. and Williamson, J.D. (1995) *Human Vaccines and Vaccination*. BIOS Scientific Publishers, Oxford.

Sheikh, N., Rajananthanan, P. and Morrow W.J.W. (1996) Immunological adjuvants: mechanisms of action and clinical applications. *Expert Opinions on Investigational Drugs*, **5**, 1079–1099.

Stott, E.J. and Schild, G.C. (1996) Strategies for AIDS vaccines. *J. Antimicrob. Chemother.*, **37** Suppl. B, 185–198.

Ulmer, J.B., Sadoff, J.C. and Liu, M.A. (1996) DNA vaccines. *Curr. Opin. Immunol.*, **8**, 531–536.

Electronic resources

For specific sources, use of a search engine such as Alta Vista or Webcrawler will provide links to relevant sites and, since URL addresses change frequently, may be more up to date than the links provided below. Some useful examples are listed below, and provide a variety of links to related material:

DNA Vaccine Web
http://www.genweb.com/Dnavax/dnavax.html

'Epidemiology and Prevention of Vaccine-Preventable Diseases', CDC course textbook
http://www.cdc.gov/nip/epvtable.htm

Global Programme for Vaccines and Immunization (GPV), WHO immunization resource
http://www.who.ch/programmes/gpv/GPV_Homepage.html

The Vaccine Page, Vaccine News & Internet Resource
http://vaccines.com/

Chapter 4

Antiviral drugs

4.1 The history of antiviral drug development

A general problem for the development of safe and effective antiviral drugs is the close similarity between viral and cellular metabolism, where many processes use the same pathways and enzymes. A great deal of effort has gone into identifying specific functions which are essential to the virus but which are either not present in or not essential to the cells of the host organism. Despite these problems, antiviral drugs are actually not a recent development. The first such drugs were derived from the sulfonamide antibiotics in the early 1950s, when thiosemicarbazone compounds were found to be active against poxviruses, raising the possibility of antiviral chemotherapy against smallpox (which was still a major cause of human disease at that time). Further efforts identified methisazone, a synthetic thiosemicarbazone, which appeared to be effective in preventing smallpox and in controlling complications of vaccination with the live vaccinia virus vaccine. Ultimately, however, vaccination proved more effective at bringing smallpox under control and, since the elimination of smallpox, vaccinia vaccine has not been in general use. However, other antiviral compounds were identified. As noted above, selectivity (see *Figure 4.1*) was a problem with many of the early antiviral compounds, and toxic side effects prevented their widespread use.

4.2 Current antiviral drugs

The number of licensed antiviral drugs has doubled in the past 3 years, and there are many more in the later stages of development. Before discussing the drugs currently available, it is worth considering the underlying principles of these drugs. *Figure 4.2* shows the stages of virus infection that may be vulnerable to antiviral intervention, together with examples of inhibitors of these stages. The broad classes into which current drugs fit are shown in *Table 4.1*. The majority of available drugs are nucleoside analogs, derived from and structurally similar to the

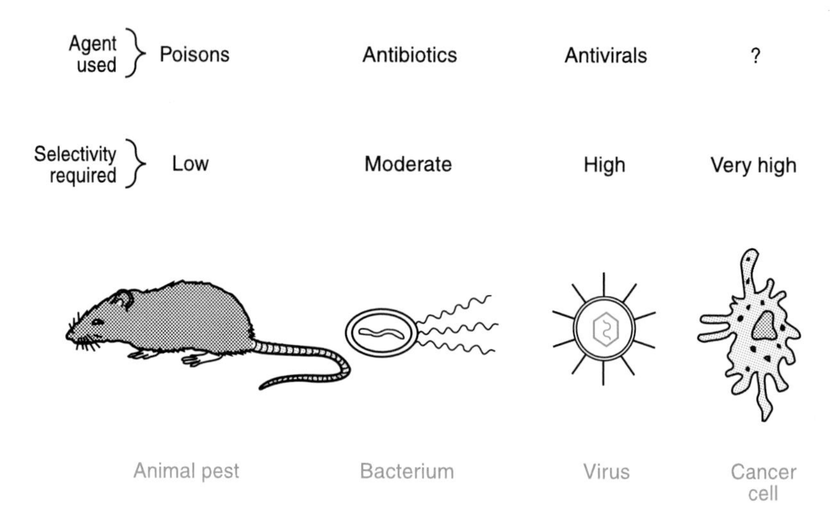

Figure 4.1: Comparison of selectivities required for antiviral drugs and other agents.

building blocks of DNA, but different enough to inhibit DNA synthesis. In many cases (including aciclovir and AZT), this involves the loss of the reactive 3′-hydroxyl group to which the next base is attached in the forming chain. With this lost, the DNA chain cannot be extended further. Other nucleoside analogs which have this group (such as famciclovir) may allow some extension, but disrupt the forming chain due their different shape.

While we are now at the point where there is some choice in which antiviral drug to use, at least for herpesviruses and the human immunodeficiency virus, it was not until the development of aciclovir [9-(2-hydroxyethoxymethyl)guanine] in the late 1970s that a non-toxic, effective antiviral drug became available. This compound represented the first of the current generation of antiviral drugs. As discussed above aciclovir is a nucleoside analog, similar to the DNA component guanosine but with an acyclic sugar group (see *Figure 4.3*). It is effective against HSV, which causes genital and oral herpes and, to a lesser extent, against VZV, which causes chickenpox and shingles. Aciclovir itself is a ‘prodrug’; it is a precursor of the active antiviral drug rather than being active against the virus itself. It is converted within the body into the active form; in this case, aciclovir triphosphate. The selectivity of aciclovir is based on

Table 4.1: Major classes of antiviral drugs in use

Drug type	Examples	Target viruses
Nucleoside analogs	Aciclovir, AZT	Herpesviruses, HIV
Cytokines	Interferon	Hepatitis B, hepatitis C, papillomaviruses
Proteinase inhibitors	Saquinavir, Ritonavir	HIV
Non-nucleoside reverse transcriptase inhibitors (NNRTIs)	Nevirapine	HIV

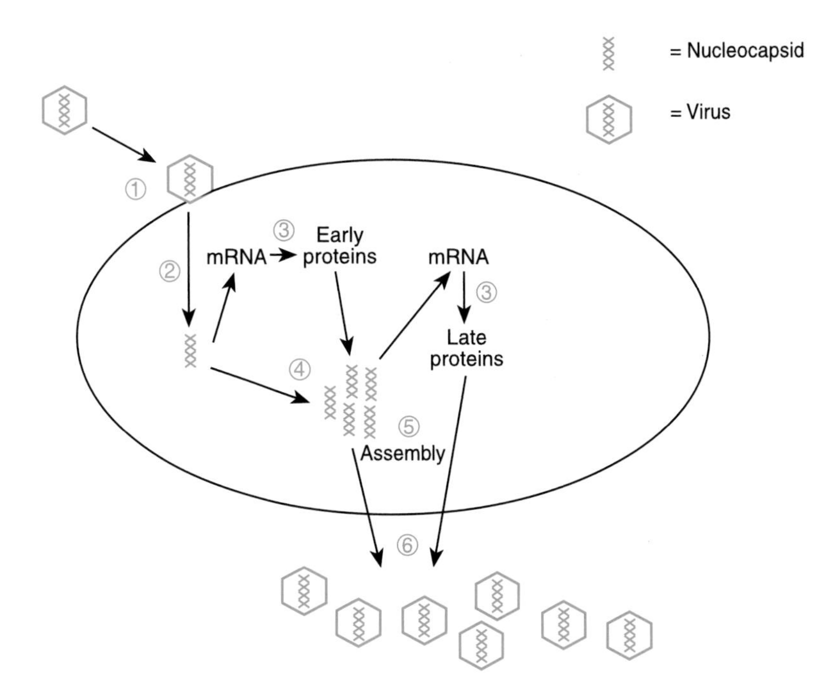

Figure 4.2: Schematic diagram of virus infection, showing the stages of viral infection vulnerable to antiviral intervention.

Diagram	*Stage*	*Examples*
1	Binding to viral receptor	Soluble CD4 (HIV); WIN compounds (picornaviruses)
2	Penetration of cell	Amantadine (influenza)
3	mRNA function	Interferon (hepatitis B and C)
4	DNA synthesis	Aciclovir (herpesviruses); AZT (retroviruses)
5	Viral assembly	Saquinavir and other proteinase inhibitors (HIV)
6	Transport and release of virus	Amantadine (some influenza strains)

Many other viral functions are being evaluated in preliminary studies. These include the production of precursors for viral metabolism, viral gene regulation, the post-translational modification of viral proteins and many others far too numerous to list here: some such approaches are discussed in Sections 4.6 and 4.7.

the presence in herpesvirus-infected cells of a viral enzyme, thymidine kinase (TK), which initiates the phosphorylation of aciclovir. Cellular kinase enzymes then continue this process, producing aciclovir triphosphate, which is used as a component of the growing DNA chain. The cellular TK is more selective and will not phosphorylate aciclovir, leaving it in the inactive prodrug form. In addition, the cellular enzyme is only produced by cells at specific stages of their life cycle, but the viral TK is produced routinely by herpes simplex viruses since it allows them to replicate in cells without waiting for the cellular enzyme to appear. The

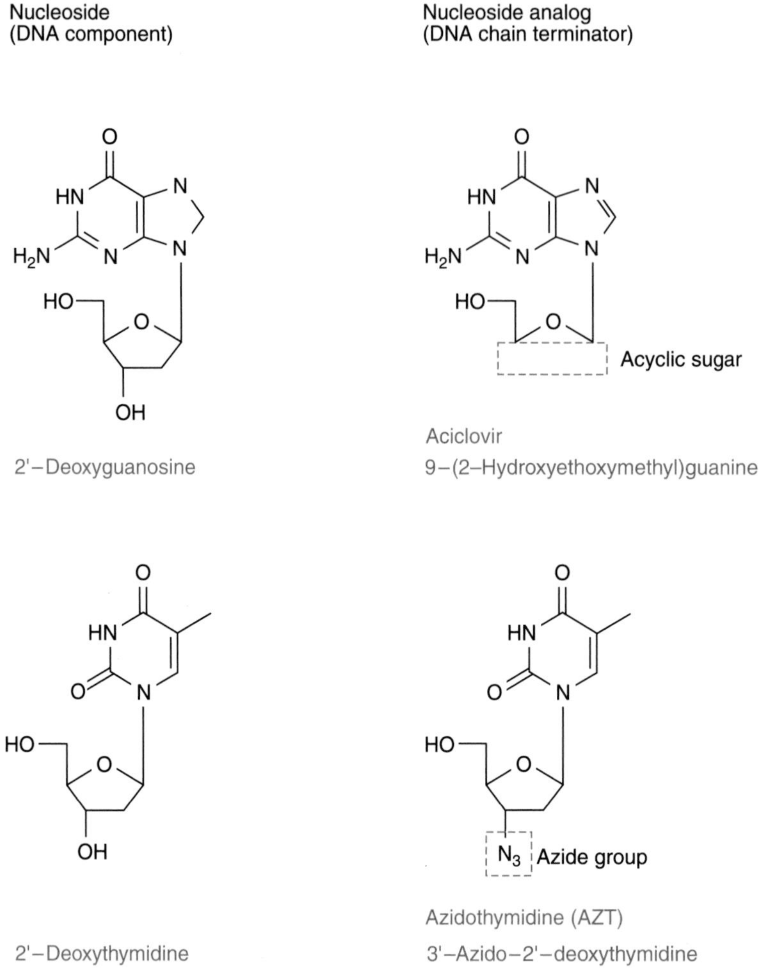

Nucleoside
(DNA component)

Nucleoside analog
(DNA chain terminator)

Acyclic sugar

2'–Deoxyguanosine

Aciclovir
9–(2–Hydroxyethoxymethyl)guanine

Azide group

2'–Deoxythymidine

Azidothymidine (AZT)
3'–Azido–2'–deoxythymidine

Figure 4.3: Nucleosides and nucleoside analogs.

activation and function of aciclovir is shown in *Figure 4.4*. As can be seen, when aciclovir is added to the DNA chain, the lack of a large part of the deoxyribose sugar means that no further bases can be added, and DNA synthesis is terminated. As a result of the requirement for viral enzymes to activate aciclovir, the drug is not toxic to uninfected cells. Additional selectivity results from a greater uptake of aciclovir by infected cells, and selection against the use of aciclovir triphosphate by the cellular (but not the viral) DNA polymerase enzyme.

Another nucleoside analog in widespread use is the anti-HIV drug AZT (3′-azido-3′-deoxythymidine, zidovudine). The structure of AZT is also shown in *Figure 4.3*. Unlike aciclovir, AZT is phosphorylated to the triphosphate (active) form by cellular kinase enzymes. However, most cellular DNA polymerases do not incorporate AZT triphosphate into the

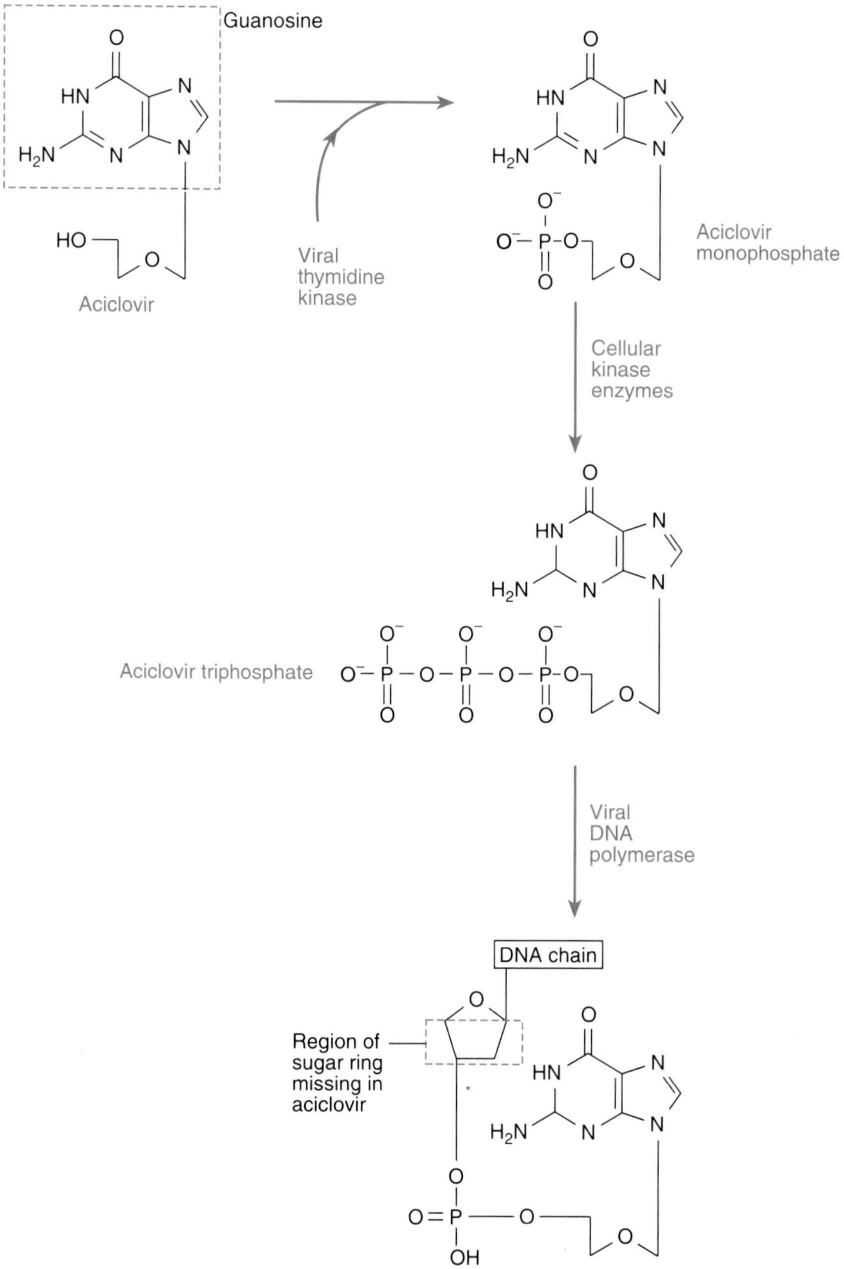

Figure 4.4: Activation of aciclovir.

growing DNA chain. The HIV reverse transcriptase enzyme is less selective, and will incorporate AZT triphosphate. The azido (N_3) group on the 3′ carbon of the deoxyribose sugar ring blocks the addition of further bases to the DNA chain, as discussed above, terminating DNA synthesis. Selectivity of AZT is not as high as that of aciclovir, and some cellular enzymes will incorporate AZT triphosphate. As a result, the toxicity of AZT is higher than that of aciclovir and is a significant problem in clinical practice. However, AZT was the first drug to be licensed for use against HIV and, in the absence of anything better, was used extensively.

Currently available antiviral drugs are listed in *Table 4.2.* Although most are nucleoside analogs as discussed, and work continues on improved nucleoside analog compounds with improved activity or availability, there is a trend away from such drugs towards new approaches, and other types of inhibitor are available. Since drug combinations work best if different aspects of virus metabolism are inhibited (see Section 4.4), the expanding use of combination therapy is a major force behind the search for non-nucleoside analog drugs. Such drugs include the non-nucleoside analog reverse transcriptase inhibitors, and proteinase inhibitors, both developed for use against HIV. While nucleoside analog reverse transcriptase inhibitors interact with the active site of the enzyme, the non-nucleoside analogs bind to another region of the reverse transcriptase. Proteinase inhibitors currently available resemble the natural substrate of the aspartyl proteinase of HIV and are related chemically to the natural peptide substrate. They act to inhibit the proteolytic cleavage of the *gag* polyprotein, which is necessary for HIV maturation. Many other functions, both virus specific and (more rarely) cellular, which are essential to the virus are being studied for possible antiviral uses. Some of these are shown in *Figure 4.2,* and are discussed here or in Sections 4.5 and 4.6.

The situation with antiviral drugs is very similar to that with antibiotics. We have now reached the stage where we have some choice in which drugs to use, and have the ability to use combinations of complementary drugs, but we are already seeing the appearance of clinically significant resistance to multiple drugs.

4.3 Resistance to antiviral drugs

One potential problem with having so few drugs available for use is the development of drug resistance by viruses. This directly parallels bacterial resistance to antibiotics, and is being studied closely at the molecular level. In a clinical setting, resistance to several antiviral drugs has been observed. Resistance to aciclovir most commonly involves mutation in conserved regions of the TK gene, resulting in complete loss of function. TK is a 'luxury function' for the virus, in that it is useful but not essential for replication. However, it is essential to initiate the phosphorylation of aciclovir (see *Figure 4.4*). Mutants lacking thymidine kinase activity (TK^-) are less pathogenic and do not appear to cause such extensive disease, at least in immunocompetent patients. However, TK^- herpes-viruses can cause problems in severely immunocompromised patients, most notably those with AIDS. TK^- mutant herpes simplex viruses are thought to exist as a small fraction (0.01–0.1%) of the virus population in

Table 4.2: Antiviral drugs in current use

Drug name	Chemical type	Virus target(s)	Effect
(a) Herpesviruses (*Herpesviridae*)			
Aciclovir (Zovirax)	Acyclic nucleoside analog [9-(2-hydroxyethoxymethyl) guanine]	HSV, VZV	DNA polymerase inhibitor (chain terminator)
Cidofovir (HPMPC, Vistide)	(*S*)-1-(3-Hydroxy-2-phosphonylmethoxypropyl) cytosine	CMV (HSV)	DNA polymerase inhibitor (competitive)
Famciclovir (Famvir)	Acyclic nucleoside analog [9-(4-acetoxy-3-acetoxy methylbut-1-yl) guanine]	VZV, HSV	DNA polymerase inhibitor (chain terminator)
Foscarnet	Pyrophosphate analog (trisodium phosphono-formate hexahydrate)	HSV, other HHVs	DNA polymerase inhibitor (non-competitive, at pyrophosphate-binding site)
Ganciclovir (Cymevene)	Acyclic nucleoside analog [9-(1,3-dihydroxy-2-propoxymethyl) guanine]	CMV, HSV	DNA polymerase inhibitor (competitive)
Idoxuridine	Deoxynucleoside analog (5-iodo-2′-deoxyuridine)	HSV	Incorporated into DNA, disrupts transcription and translation
Trifluorothymidine	(Nucleoside analog)	HSV	Incorporated into DNA, disrupts transcription and translation
Valaciclovir (Valtrex)	L-Valyl ester of aciclovir	VZV	As aciclovir
Vidarabine (adenosine arabinoside, ara-A)	Nucleoside analog (9-β-D-arabinofuranosyl adenine monohydrate)	HSV, VZV	DNA synthesis inhibitor (unknown method)
(b) Human immunodeficiency virus (*Retroviridae*)			
Didanosine (Videx, DDI)	Dideoxynucleoside analog (2′,3′-dideoxyinosine)	HIV	Reverse transcriptase inhibitor (competitive and chain terminator)
Idinavir (Crixivan)	Peptide mimetic[a]	HIV	Disrupts protein cleavage during virus maturation
Lamivudine (Epivir, 3TC)	Dideoxynucleoside analog [2′,3′-dideoxy-3′-thiacytidine, (−) enantiomer]	HIV	Reverse transcriptase inhibitor (chain terminator)
Nelfinavir (Viracept)	Peptide mimetic[a]	HIV	Disrupts protein cleavage during virus maturation
Nevirapine (Viramune)	Non-nucleoside reverse transcriptase inhibitor (11-cyclopropyl-5,11-dihydro-4-methyl-6H-dipyrido[3,2-b:2′,3′-e][1,4]diazepin-6-one)	HIV	Reverse transcriptase inhibitor (binds outside active site)
Ritonavir (Norvir)	Peptide mimetic[a]	HIV	Disrupts protein cleavage during virus maturation

(*continued over page*)

Table 4.2: *Continued*

Drug name	Chemical type	Virus target(s)	Effect
Saquinavir (Invirase)	Peptide mimetic[a]	HIV	Disrupts protein cleavage during virus maturation
Stavudine (Zerit, D4T)	Deoxynucloside analog (2′,3′-didehydro-2′-deoxythymidine)	HIV	Reverse transcriptase inhibitor (chain terminator)
Zalcitabine (Hivid, DDC)	Dideoxy nucleoside analog (2′,3′-dideoxycytidine)	HIV	Reverse transcriptase inhibitor (competitive and chain terminator)
Zidovudine (Retrovir)	Azido-deoxynucleoside analog (3′-azido-3′-deoxythymidine)	HIV	Reverse transcriptase inhibitor (chain terminator)
(c) Myxoviruses (*Orthomyxoviridae, Paramyxoviridae*)			
Amantadine	Primary amine 'cage' (tricyclo[3,3,1^{3,7}]dec-1-ylamine hydrochloride)	Influenza A virus (*Orthomyxoviridae*)	Interferes with viral penetration and uncoating, may interfere with assembly
Ribavirin[b]	Nucleoside derivative (1-β-D-ribofuranosyl-1,2,4 triazole 3-carboxamide)	RSV (*Paramyxoviridae*) Other viruses inhibited *in vitro*	Several proposed, none certain
Rimantadine (α-methyl-amantidine)	Primary amine 'cage' (α-methyl-tricyclo [3,3,1^{3,7}]dec-1-ylamine hydrochloride)	Influenza A virus (*Orthomyxoviridae*)	Interferes with viral penetration and uncoating, may interfere with assembly
(d) Other viruses			
Alpha-interferon	Recombinant protein	Hepatitis B virus (*Hepadnaviridae*) Hepatitis C virus (*Flaviviridae*) Papillomavirus (*Papovaviridae*)	Modulation of immune response, cellular antiviral effects

[a]Complex structure derived from natural peptide substrate (VSQNYPIV).
[b]Also active against Lassa fever virus (*Arenaviridae*).

an infected individual, since they are at a replicative disadvantage compared to TK^+ strains. By inhibiting the replication of viruses with a functional TK, aciclovir will select for such mutants rather than requiring *de novo* mutation. Other, less common, aciclovir-resistant mutants may have increased TK substrate specificity (preventing aciclovir phosphorylation) or mutations in the DNA polymerase which prevent aciclovir incorporation (although these cannot involve complete loss of polymerase function since this is lethal to the virus). DNA polymerase mutants can also be resistant to the non-nucleoside polymerase inhibitor foscarnet, which can be a problem since this is a main drug of choice for use against aciclovir-resistant strains. There is also some evidence for changes in other viral proteins being involved in aciclovir resistance. Mutations giving a TK^+ aciclovir-resistant phenotype do not appear to occur frequently *in vivo*, possibly because they require very specific mutations to avoid loss of TK function. However, such mutations result in more pathogenic viruses, since TK function is retained, which may be a problem in the future. Similar patterns of resistance have been observed with VZV and with CMV, where resistance involves either a failure to phosphorylate the inhibitor used or alterations in the DNA polymerase.

Resistance to anti-HIV drugs is a significant problem in clinical practice. Firstly, HIV mutates very rapidly. Secondly, patients with late-stage HIV infection are by definition immunosuppressed, and it is known from work with antiherpesvirus drugs that immunosuppression favors the development of drug-resistant mutant viruses. Resistance to AZT appears to result from single base changes at one of at least four specific sites within the viral reverse transcriptase gene. As with herpesvirus DNA polymerase mutants, loss of enzyme function is lethal, so the mutant enzymes are still functional. Similar resistance has been observed to other nucleoside analog reverse transcriptase inhibitors such as dideoxyinosine. Early evidence that this involves mutations at different sites and that the effect of these mutations may mask AZT resistance is countered by the recent appearance of multiply resistant viruses.

Despite promising results *in vitro*, the development of resistance to non-nucleoside analog reverse transcriptase inhibitors of HIV has often been very rapid, apparently because these compounds do not bind to an essential site on the enzyme. This means that mutations producing resistance to the drug can arise without significant adverse effects on virus replication, and as a result they occur far more readily. Despite this, one such drug (nevirapine) has been licensed, primarily for use in combination with other drugs where the development of resistance is suppressed (see Section 4.4).

4.4 Combination therapy

The use of multiple drugs sequentially is now generally thought to be of limited benefit, but the use of multiple drugs simultaneously ('combination therapy') may provide significant benefits for two reasons. Since any mutant would have to develop resistance to two (or more) drugs simultaneously, this is less likely, particularly for HIV as noted above. In addition, the effects of some drugs may combine to enhance their antiviral effect. This is referred to as a 'synergistic' effect, with the combined effect being greater than would be expected by simply adding the individual effects of the drugs together, and differs from additive or antagonistic effects which may be seen with other drug combinations (*Figure 4.5*). Such effects are also seen with combined antibiotic therapy

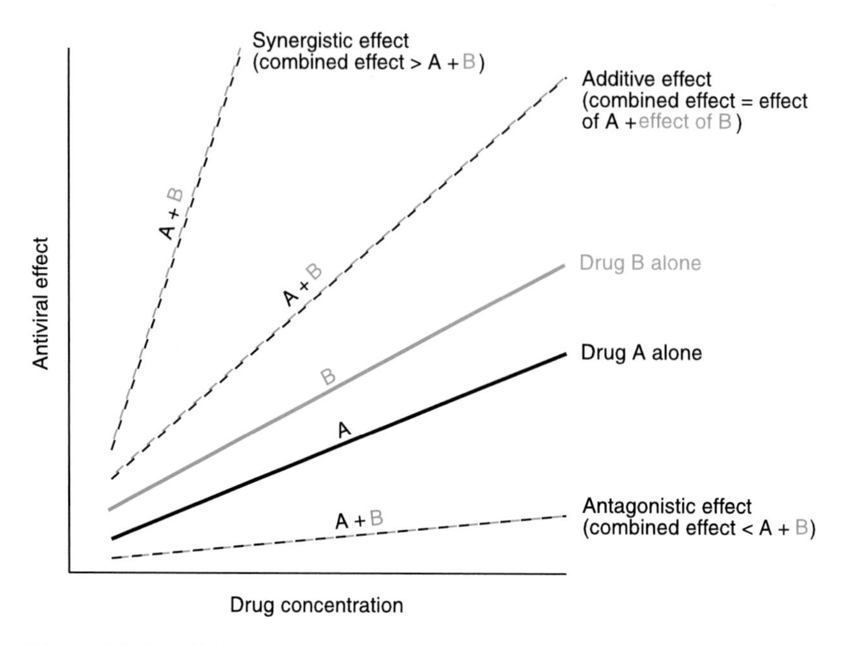

Figure 4.5: Possible outcomes of drug combinations.

against bacterial diseases. The molecular basis of synergy with antiviral drugs is poorly understood at present, and synergistic combinations are normally derived from extensive testing programs where multiple drugs are tested singly and in combinations at various levels. Efforts to establish a rational basis for combinations are under way, since random use of combined therapy could favor the emergence of resistant mutants. Combination therapies of course rely on the availability of multiple drugs approved for use in humans, ideally with different modes of action, such as the use of nucleoside analog DNA synthesis inhibitors and proteinase inhibitors which block virus maturation, used in combination against HIV. Such combinations are only now becoming possible. Many different combinations have been evaluated, including AZT with interferon, proteinase inhibitors, antiherpesvirus or antifungal drugs. Some of these, for example α-interferon and the antiherpesvirus drug foscarnet, have licenses for (other) uses in humans and may represent possible therapies.

Reduction of the dose of a drug resulting from its use in combination therapy can reduce drug-related toxicity, which is a problem with many antiviral drugs, notably AZT. A range of combined therapies are already in clinical trials, based mostly around combinations including well characterized drugs such as AZT or aciclovir. However, the most interest surrounds the combination of nucleoside analogs and proteinase inhibitors in HIV disease. In some cases, it has been reported that such combinations may represent a 'cure for AIDS'. Such optimism is very likely to be premature, since the virus can survive in an inactive (proviral) state (see Section 1.4.5) where it is immune to the effects of these drugs.

Since they exert their effects on replicating virus, inactive virus cannot be cleared by their use. While it may be possible to continue combination therapy for long enough for any inactive virus to reactivate and thus become vulnerable, this has yet to be proved. In addition, the drug regimens used are highly complex, requiring excellent patient compliance to avoid underdosing and the possible development of resistance. They are also extremely expensive ($15 000–$20 000 per year), and have significant side effects that prevent their proper use, in some cases by more than 50% of patients in a study group. Thus, while combination therapy represents a significant step forward, we are not yet in a position where we can realistically talk of a cure, even for a virus such as HIV where multiple antiviral drugs are available.

4.5 Interferons

The nature and effects of interferons have been summarized in Section 2.1. Interferon was first reported in 1957 as a soluble agent produced by cells treated with inactivated influenza virus which was capable of inhibiting infection of cells from hen's eggs with live influenza virus. This direct antiviral effect raised the hope that interferons would prove to be effective antiviral agents. Studies were carried out using purified interferons, but interferon is naturally produced in very small amounts and it was extremely difficult and expensive to obtain sufficient quantities. When cloning and expression of interferon genes became possible, large-scale studies could be carried out.

It soon became clear that interferons were not the 'magic bullet' that had been hoped for. Studies with respiratory viruses showed that interferon was effective if administered before infection. Unfortunately, the only way in which such administration would be possible outside the laboratory would be continuous dosing, and this would be unproductive since interferon itself irritates and damages the nasal epithelium, producing symptoms similar to those of viral infection. Knowing as we do now that interferon is actually a potent immunoregulatory agent, this is not surprising. The idea that interferon was a simple antiviral agent has been replaced by our current awareness of its role as part of the complex immunological network. Unsurprisingly, adding large amounts of one component (interferon) to such a system produces complex and potentially damaging effects. Following these disappointing results, much of the interest in interferon as an antiviral agent was lost, although α-interferon has been used successfully in the treatment of a range of cancers.

Recently, interest in the antiviral use of α-interferon has been revived by promising results against chronic infection with hepatitis B and C viruses. While interferon therapy is not effective in all patients, a substantial proportion do respond favorably, particularly those with chronic hepatitis B virus infection. It may well be that interferon will prove valuable in specific applications, but it is clear that early hopes for the 'magic bullet' were unfounded.

4.6 Development of antiviral drugs

Almost every area of viral metabolism is under investigation somewhere as a possible approach to antiviral drug development. Many investigations are simply the application of approaches known to work with one type of virus to another. A prime example of this is the development of proteinase inhibitors against herpesviruses and enteroviruses following the apparent success of this approach for HIV. A more limited example is the development of modified nucleoside analogs with improved activity against those herpesviruses (CMV, VZV, EBV) less inhibited by currently available drugs. In the case of VZV, two very promising candidates are shown in *Figure 4.6*. In an illustration of the problems of drug development, both have run into severe problems. 5-Propynyl-ara-U failed long-term toxicity assays and has been withdrawn from testing. With Brovavir, the drug inhibits uracil metabolism. While this is not normally a problem, some of the patients treated with Brovavir in a

5-Propynyl-ara-U: 1-[β-D-arabinofuranosyl]-5 propynyluracil (882C)

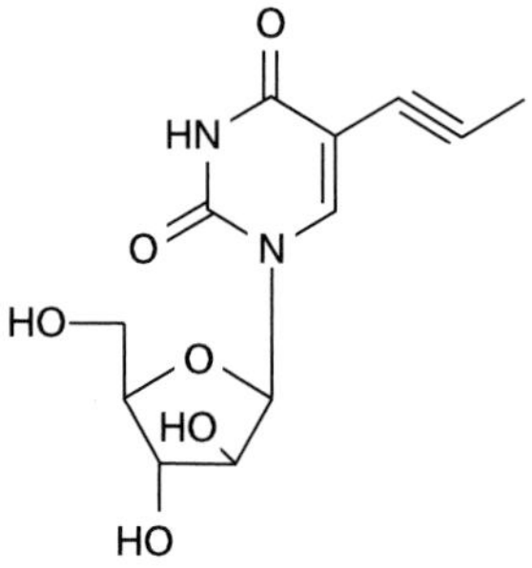

Brovavir: 1-β-D-arabinofuranosyl-E-5-[2-bromovinyl]uracil (BVaraU, Sorivudine)

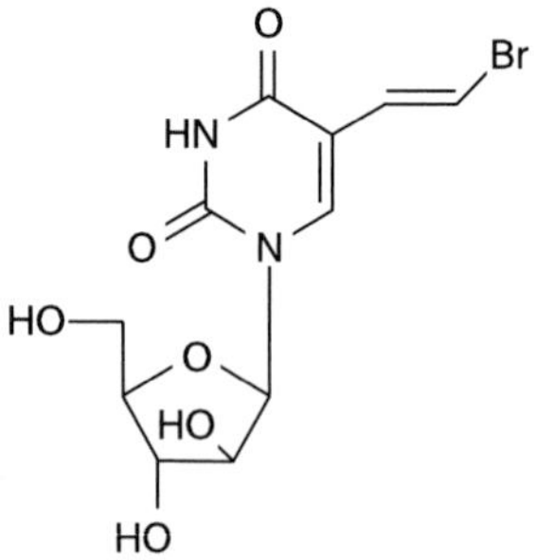

Figure 4.6: Examples of more active nucleoside analog-based drugs for use against VZV. 5-Propynyl-ara-U: 10 times more active against VZV than aciclovir, less active against other herpesviruses, but has been withdrawn due to toxicity problems. Brovavir: 10 *thousand* times more active against VZV than aciclovir *in vitro*, less active against other herpesviruses, can cause severe toxicity if co-administered with uracil-based drugs.

Japanese trial and immediately following the issue of a limited license were also being treated with 5-fluorouracil (an anticancer agent), and the toxic levels of 5-fluorouracil which built up resulted in the deaths of at least 15 patients. This situation would not normally occur where prescribing doctors were aware of other medication, but needless to say this has resulted in a major slowing down of Brovavir development.

Another approach to enhancing drug efficacy is to attempt to modify the chemical structure to get larger amounts of the orally administered drug into the bloodstream (to 'increase oral bioavailability'); examples active against herpesviruses include valaciclovir and famciclovir (*Figure 4.7*). With valaciclovir (now marketed under the name Valtrex), an

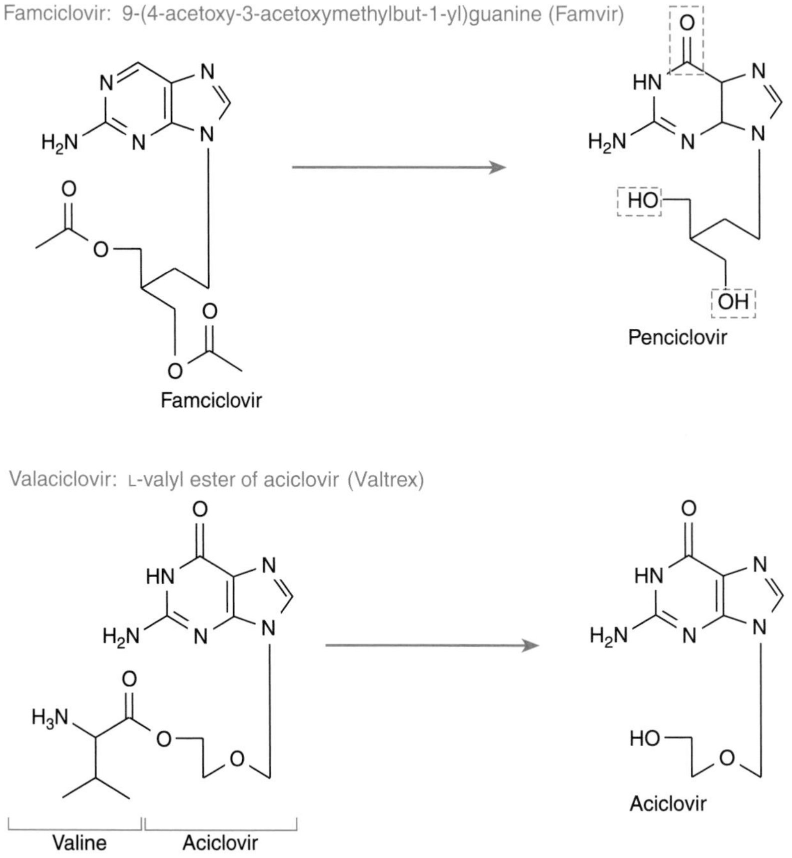

Figure 4.7: Examples of nucleoside analog-based drugs with improved oral bioavailability for use against herpesviruses. Famciclovir: fivefold higher oral bioavailability (serum levels relative to the dose of drug given orally) than aciclovir. Famciclovir is a prodrug and is converted to the active form (penciclovir) within the body. The structure of penciclovir is shown, and structural changes from famciclovir are outlined by dotted lines. Valaciclovir: three- to fourfold higher oral bioavailability than aciclovir, valine residue cleaved within body to produce aciclovir.

attached valine residue greatly enhances oral bioavailability, but is removed in the process so that the effective antiviral drug is actually aciclovir. Valaciclovir is a prodrug (precursor form, converted to the active drug) of aciclovir, which is itself a prodrug of aciclovir triphosphate. In another instance, famciclovir is also a prodrug, and is converted to a related, active drug (penciclovir). Penciclovir is poorly absorbed by the oral route, while famciclovir has good oral bioavailability and is readily converted to penciclovir and then to penciclovir triphosphate within the body.

It should be noted that even a slight modification can exert major effects on the activity of a drug. Both penciclovir and ganciclovir are closely related to aciclovir (see *Figure 4.8*), and yet they have substantially different properties. Aciclovir has very low toxicity and is active against HSV-1 and -2 and VZV. Ganciclovir is significantly more toxic than aciclovir, but is also effective against CMV. Penciclovir has low toxicity and is active against the same viruses as aciclovir, but is also active against the completely unrelated hepatitis B virus, apparently by the activity of cellular enzymes since this virus does not produce an enzyme able to phosphorylate the drug. The intracellular metabolism of aciclovir and penciclovir are also quite different in other ways. As this would suggest, in compound testing it is common to see major, unpredictable changes in activity and toxicity from even very minor chemical changes, reflecting the complexity of biological systems.

The problems of developing antibacterial agents are also seen with antiviral drugs. It is relatively easy to inhibit the function of a specific viral enzyme in the test tube, but far more difficult in cells or whole animal (or human) systems. In addition, many of the inhibitors will also kill both infected and uninfected cells. Still more will be toxic in animals, for example by their effects on specific specialized cell types. Some of those

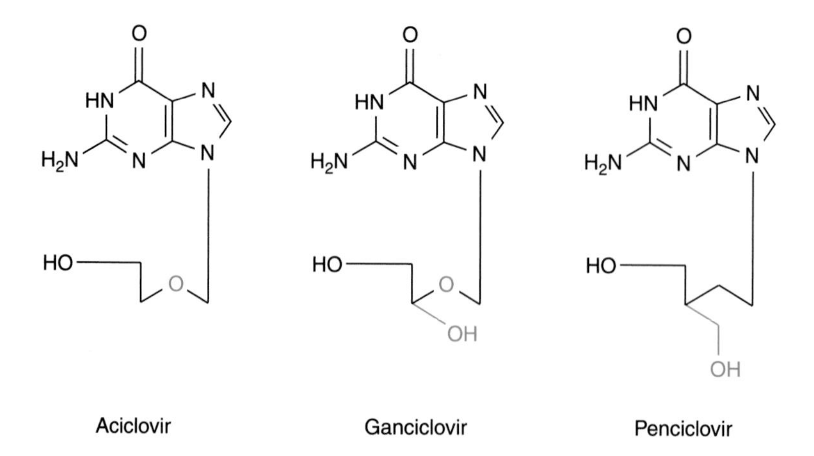

Figure 4.8: Comparative structures of antiherpesvirus nucleoside analogs.

which have reached this stage of testing will then prove to be toxic in humans. Thus, of the many compounds active against viruses in initial laboratory assays, only a very few will even reach the stage of clinical testing, and even fewer will be licensed for general use. As with everything else, it is the early identification of these 'winners' which is the problem facing pharamaceutical companies (*Figure 4.9*).

There are two basic approaches to the development of antiviral drugs. These are screening of compound libraries, or rational design of molecules likely to have the desired properties or effects. Screening uses relatively simple tests to assay the antiviral effects of newly synthesized compounds, compounds obtained from biological sources or compound libraries maintained by pharmaceutical companies. Compounds are screened whether or not they are thought likely to inhibit viral replication. Such methods have been used extensively in the past, when molecular mechanisms were less well understood, but the emphasis now is switching towards characterizing the molecular events underlying a critical aspect of viral replication and designing compounds likely to inhibit those events. An excellent example of this latter approach was the design of inhibitors of the HIV proteinase.

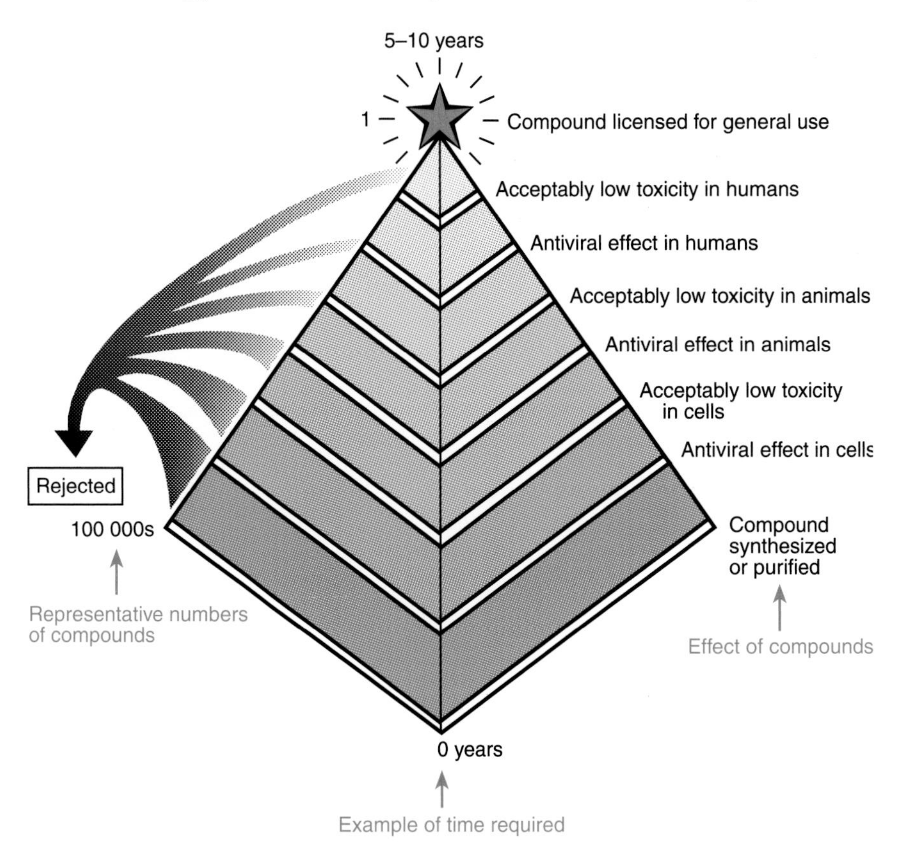

Figure 4.9: The pyramid of drug discovery.

In molecular studies of HIV replication, many aspects of virus metabolism were characterized. Among these was a proteinase function essential for the formation of mature (infectious) virus particles (see Section 1.4.5). The proteinase was identified as an unusual enzyme since, unlike cellular proteinases, it appeared to rely on the presence in the active site of the enzyme of an aspartic acid residue. The lack of such a proteinase in normal cells meant that this presented a promising target for the development of an antiviral drug; inhibition of its function should not interfere with cellular metabolism. Inhibitors based on the target peptide cleaved by the HIV proteinase were developed, and these were then refined by chemical modifications giving known alterations in the properties of the inhibitors until a highly active and specific inhibitor was produced. The relative structures of the natural peptide substrate of the HIV-1 proteinase and a peptide-based proteinase inhibitor are shown in *Figure 4.10*. The modifications were selected to make this inhibitor interact very efficiently with the active site of the viral proteinase and to inhibit its function as a result of this high avidity. This required extensive

(a) Natural substrate of the HIV-1 proteinase:

Cleavage site

Valine Serine Glutamine Asparagine Tyrosine Proline Isoleucine Valine

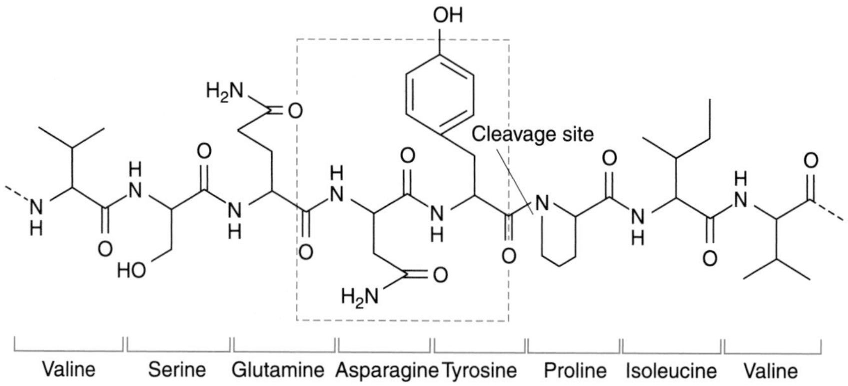

(b) Proteinase inhibitor Saquinavir / Invirase:

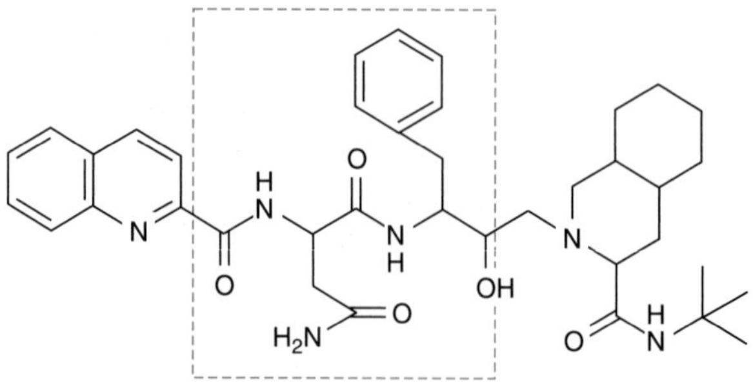

Figure 4.10: Structures of (a) the natural substrate of the HIV-1 proteinase and (b) the proteinase inhibitor Saquinavir/Invirase (note similarity of regions in dotted box).

knowledge of the enzymic and structural properties of the proteinase. Some of the modifications made to such compounds are also intended to maximize uptake by the target cell and to minimize degradation both during uptake and within the cell. This same approach is now being applied to other viruses which produce virus-specific proteinases, notably herpesviruses and picornaviruses (see Sections 1.4.4 and 1.4.5), and represents an excellent example of rational drug design.

Underlying rational drug design is an understanding of the molecular interactions of drug and virus at the sub-molecular level. This is only possible because of recent developments in computer modeling of the structures of proteins and other biological macromolecules, which allow the interaction of inhibitors with virus to be calculated precisely. For example, the design of one class of inhibitors of rhinoviruses is based around very specific tailoring of small molecules which fit into the receptor-binding 'canyon' known from crystallographic studies to be present on the surface of the virus. These then prevent binding to the viral receptor. Although the initial inhibitors with this effect were developed by traditional methods, improved and more effective compounds have been (and are being) produced through rational design. Rational drug design and molecular modeling are rapidly developing, and underlie much of the current work on drug development. Modeling of resistant viral enzymes will also allow better understanding of the actual mechanisms involved in drug resistance and design of inhibitors to bypass such resistance. The computer has become increasingly important in the initial stages of drug design, although testing in biological systems remains as an essential next step. The complexity of biological systems can result in unexpected effects from even minor changes, as noted above, and computer-based modeling systems are very unlikely to be able to predict such effects reliably for some time to come.

4.7 Nucleic acid-based approaches to antiviral drug development

Such approaches are the subject of a great deal of attention, since (unlike the drugs discussed in Sections 4.2 and 4.6) nucleic acid 'drugs' need not rely on disruption of active viral metabolism. As a result, they could allow the targeting even of inactive viruses. The main approaches under investigation are summarized in *Figure 4.11*, and are discussed below.

One major area of development is based on 'antisense strategies'. The basic concept underlying this approach is that of introducing an 'antisense' RNA complementary to a viral RNA into an infected cell, where it will form a dsRNA duplex by base pairing with the viral genomic or messenger RNA. This can have antiviral effects in a number of ways. Simply by locking up the viral RNA in a duplex, its function can be inhibited. Alternatively, the RNA duplex may be digested by dsRNA-specific cellular nucleases. In addition, dsRNA is a potent inducer of a

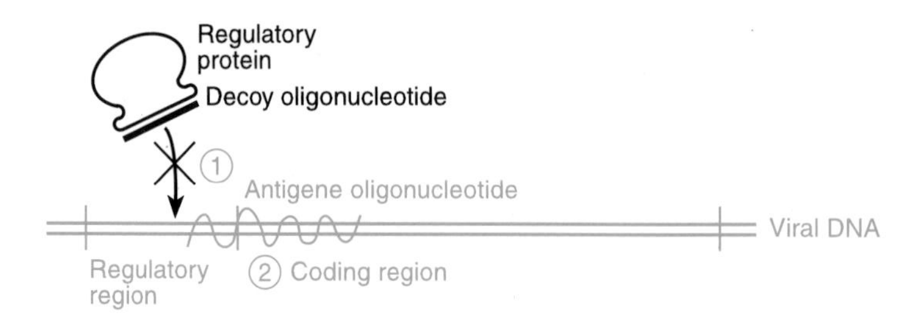

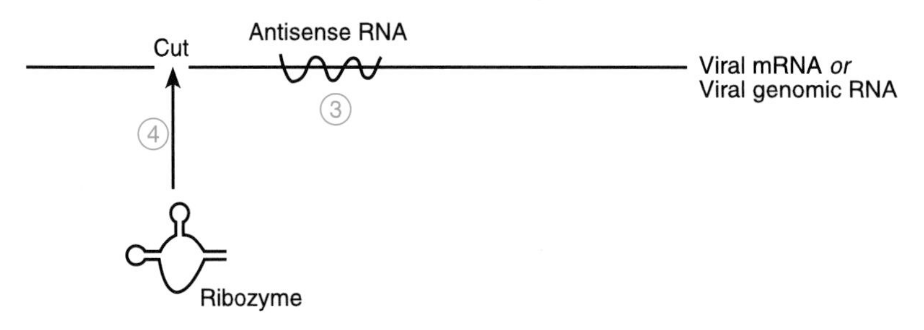

Figure 4.11: Nucleic acid-based approaches to antiviral drug development: (1) decoy, (2) antigene, (3) antisense, (4) ribozyme.

localized interferon response, which can induce antiviral effects in the cell. The antisense RNA can be introduced as an oligonucleotide or produced from a synthetic gene introduced into the cell. Use of oligonucleotides is not simple, since such molecules are readily degraded and are taken up poorly by the cell. However, chemical modifications such as the phosphorothioate modification, where one of the oxygen atoms in the phosphate groups of the oligonucleotide is replaced with sulfur, can stabilize the oligonucleotide against enzymic degradation and increase intracellular levels. Modified oligonucleotides are already in trials for use as antiviral agents. Oligonucleotides are highly complex, and to obtain specific oligonucleotides for use as probes or primers [which are used, for example, in the polymerase chain reaction (PCR), discussed in Section 6.2.3] is highly expensive. Despite this, it is anticipated that economies of scale could make the production of even such a complex antiviral drug cost-effective.

A related approach is the introduction of 'antigene' (rather than antisense) oligonucleotides. These bind to the DNA itself, forming a short region where the classical 'double helix' of DNA is replaced by a 'triple helix', which can be stabilized by the presence of abnormal bases in the oligonucleotide or the chemical linkage of intercalating agents (which slide between the chains of a double helix) to the molecule. Such a structure appears to inhibit gene transcription as well as interactions with regulatory proteins in the region of the triple helix. While effects have been apparent in some systems with this strategy, they remain highly experimental.

Introducing a synthetic gene is effectively a limited form of gene therapy, and requires the same approaches: a method of getting the gene into the cell and of stabilizing and expressing the gene when it is inside (see Section 5.6). There are as yet no clinically accepted methods for introduction and expression of such a gene, and antiviral applications of gene therapy are in their infancy. However, this is likely to change as the many systems under development move out of trials or the laboratory and into the clinic. One potentially useful approach, intended to minimize possible toxic effects, is to put the antisense gene under the control of specific viral promoters which will only become active when other viral gene products are present. This means that the antisense RNA is not produced unless the cell is actually infected by the 'correct' virus.

There are a number of other approaches to antiviral drug design that rely on the introduction of genes or RNAs into cells. One promising approach is the production from an introduced gene of short 'decoy' RNAs corresponding to regulatory regions of the HIV genome or transcripts from it, which bind the limited supply of regulatory proteins and prevent their function. Such regions include the TAR (*trans*-activation response) and RRE (Rev-response) elements. The viral *trans*-activating protein Tat binds to the TAR region, preventing termination of transcription, while binding of the viral Rev protein to the RRE allows mRNA transport. Both of these functions are essential for viral replication, and disruption exerts a strong antiviral effect.

One of the most interesting nucleic acid-based approaches is that of *ribozymes*. The name derives from combining ribonucleic acid and enzyme. From work first published in 1981 with the RNAs of the protozoan *Tetrahymena*, it became apparent that some RNA molecules appeared to be able to cleave themselves at specific sites. A great deal of work has been performed using the RNA of viroids, sub-viral pathogens of plants which consist only of intricately folded RNA which can cleave itself – a useful function since viroids do not produce any proteins (see Section 1.8.3). Ribozyme cleavage can involve one RNA molecule cutting itself, or can be the cutting of one RNA within a structure containing two RNA chains. In the case of viroids and many other ribozymes, it relies on folding of the RNA into a 'hammerhead' (see *Figure 4.12*) which brings reactive groups into proximity causing cleavage at a specific site, although other structures (notably the 'hairpin' ribozyme) are known and are under investigation. Cleavage by ribozymes occurs at precise sequences (due to the requirement for matching sequences of bases), and it is now possible to design synthetic ribozymes which will cut at a sequence specific to the genes of a target virus, giving a very high selectivity. While ribozymes appear to be limited to RNA (owing to the lack of the reactive 2′-hydroxyl group in DNA) they are able to interfere with any RNA stage of viral replication. In addition, a ribozyme capable of catalyzing the formation of a peptide bond is now known, significantly extending the range of activities available. It has been suggested that ribozymes are a 'fossil' of an

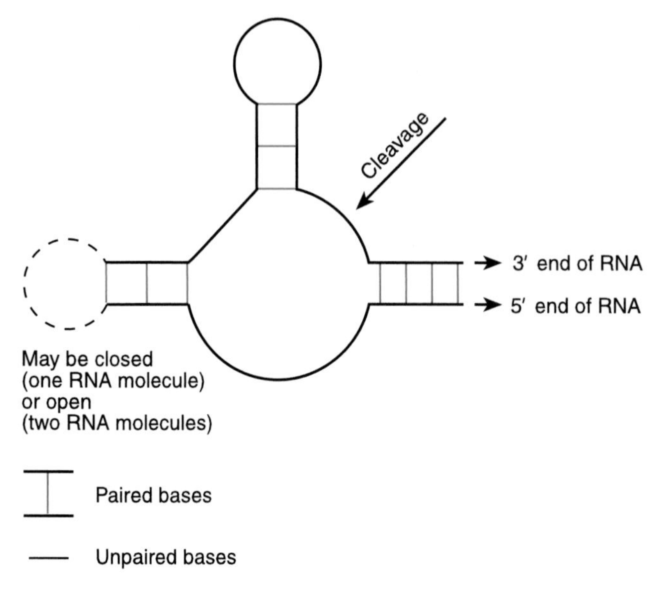

Figure 4.12: General structure of an RNA 'hammerhead'.

RNA-only biochemistry, and such functions may be the means by which proteins were first formed. Many viruses have RNA genomes, and ribozymes can inactivate the viral genome itself. Even DNA viruses must use mRNA intermediates, which can be targeted by ribozymes. Sadly, and despite much excitement when ribozymes were first discovered, practical antiviral applications are still distant.

Targeting of latent viruses by some nucleic acid-based strategies is possible, since an introduced gene could be within the cell and able to act directly on the latent DNA, and also to exert an antiviral effect when reactivation of latent virus occurs. All of the problems associated with gene therapy also apply to such approaches, including the need for targeting individual cell types and for repeated treatments due to the death of the treated cells. While it is theoretically possible to introduce 'antiviral genes' into the human germ cells and thus protect for life, this is no more than a distant prospect, and is something of an ethical minefield. The principles and practices of gene therapy are discussed in more detail in Section 5.6.

Further reading

Blair, E., Darby, G., Gough, G., Littler, E., Rowlands, D. and Tisdale, M. (1998) *Antiviral Therapy*. BIOS Scientific Publishers, Oxford.

Cohen, J. (1993) Can combination therapy overcome drug resistance? *Science*, **260,** 1258.

Collins, P. and Darby, G. (1991) Laboratory studies of herpes simplex virus strains resistant to aciclovir. *Rev. Med. Virol.*, **1,** 19–28.

De Clercq, E. (1995) Trends in the development of new antiviral agents for the chemotherapy of infections caused by herpesviruses and retroviruses. *Rev. Med. Virol.*, **5,** 149–164.

Jeffries, D.J. and De Clercq, E. (1995) *Antiviral Chemotherapy*. John Wiley & Sons, Chichester.

Jeffries, D.J. and Pauwels, R. (1993) Antivirals. In *Virology Labfax* (D.R. Harper, Ed.). BIOS Scientific Publishers, Oxford, pp. 215–230.

Symons, R.H. (1990) Self-cleavage of RNA in the replication of viroids and virusoids. *Semin. Virol.*, **1,** 117–126.

Tudor-Williams, G. and Emery, V.C. (1992) Debate: development of *in vitro* resistance to zidovudine is likely to be clinically significant? *Rev. Med. Virol.*, **2,** 123–129.

Weintraub, H.M. (1990) Antisense RNA and DNA. *Sci. Am.*, **262,** 40–46.

Electronic resources

For specific sources, use of a search engine such as Alta Vista or Webcrawler will provide links to relevant sites and, since URL addresses change frequently, may be more up to date than the links provided below. Some useful examples are listed below, and provide a variety of links to related material:

Project Inform Antiviral Treatment Information, AIDS oriented information
http://www.projinf.org/antivirals

PharmInfoNet, pharmaceutical information network drug database, includes antivirals
http://pharminfo.com/drugdb/db_mnu.html

Chapter 5

Cloning and gene therapy

with P.R. Kinchington

5.1 Viruses and cloning

A wide variety of systems have been used to allow the expression of foreign genes in prokaryotic or eukaryotic cells. While many of these are based on genetic elements that can replicate within a host cell (plasmids or episomes), viruses have been used for this purpose for a number of reasons. These include the presence of high efficiency promoters and also the ease with which a virus can introduce foreign nucleic acid into cells. While systems for the introduction of naked nucleic acid into cells are well defined, they can be difficult to use and of very low efficiency with some cell types. Viruses, optimized for this purpose by millions of years of evolution, provide a useful (if complex) alternative. Some viruses, usually those with larger genomes such as pox or herpes viruses, may be capable of independent replication even when large amounts of foreign DNA are inserted into their genome. Viruses with small genomes may require replacement of essential parts of their genome in order to carry a useful amount of foreign DNA. A good example of this is those retroviruses where replacement of essential viral genes with cellular DNA occurs during natural infections (see Section 1.4.5), and in which the insertion of cloned DNA usually requires the loss of essential genes. Viruses with inserts are maintained using another source to supply the missing functions. This can be a complete, co-infecting 'helper' virus, a modified helper virus which can supply the missing functions but not replicate itself (a 'suicide' virus), or viral genes stably expressed in the cell used to culture the virus. However, many viruses damage or kill the cells they infect. As a result, many approaches use such defective viruses, which in some systems contain little viral nucleic acid other than packaging signals.

The use of viral vectors in vaccine production is covered in Sections 3.1 and 3.3, and this chapter will concentrate on the mechanics of cloning in viral systems.

5.2 Construction of a plasmid vector

Owing to the relative simplicity and ease of use of prokaryotic systems, initial cloning of viral genes will almost always use a plasmid vector capable of replicating in bacteria. A plasmid is a covalently closed circular DNA which can replicate within a host cell. A plasmid vector is a hybrid DNA molecule derived from such a plasmid which is able to replicate and also to promote the expression of genes presented in DNA inserted into a specific site. While plasmids are relatively common in bacteria and early plasmid vectors were derived from naturally occurring plasmids, most current plasmid vectors are substantially artificial, bringing together regions of DNA from different sources (including synthetic DNA) to optimize them for particular functions.

A basic plasmid vector for cloning in prokaryotic cells is shown in *Figure 5.1*. It should be noted that plasmids for specialist applications may differ widely from those shown. Some typical components of a plasmid vector are given below.

- Promoter. Drives expression (transcription) of DNA downstream from its own location by interaction with transcription factors and RNA polymerases. Control sequences may also be present to interact with transcription factors and activators and so enhance promoter activity.

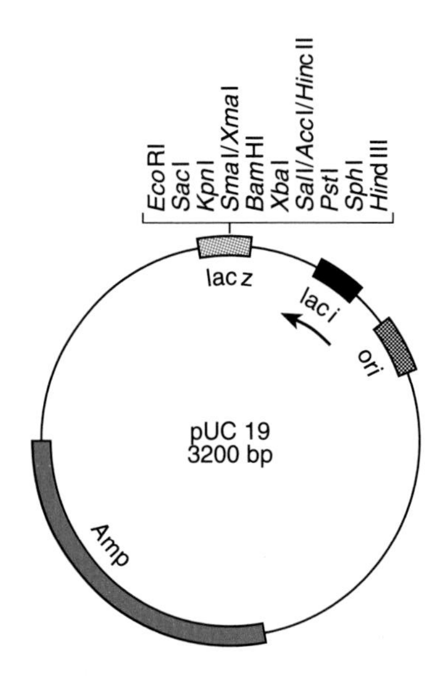

Figure 5.1: A prokaryotic cloning vector. The 'general purpose' vector pUC 19 showing:

Promoter (lac i)
Origin of replication (ori)
Indicator gene (lac z)
Selectable gene (amp)
Polycloning site (└┬┘) with the restriction enzymes cutting at that site listed.

Note that this plasmid does not contain a filamentous bacteriophage origin of replication (fl ori, used for mutagenesis).

- Cloning site. A region of synthetic DNA containing one or more sites which can be cut by different sequence-specific endonucleases (restriction enzymes) allowing foreign DNA prepared with these enzymes to be inserted at this point.
- Indicator gene. A gene whose expression can be detected by a simple assay system (β-galactosidase is often used), positioned such that its transcription is disrupted when DNA is inserted into the cloning site.
- Selectable gene. Imparts a selectable characteristic (such as antibiotic resistance, often to ampicillin) to ensure that only bacteria containing the plasmid will grow under the conditions used. May be the same as the indicator gene in some cases, with loss of antibiotic resistance as indicator.
- Origin of replication. Allows the plasmid to replicate. Different origins of replication will maintain the plasmid at anything from one copy per cell to several hundred copies per cell. Most plasmids now also contain an origin of replication from single-stranded DNA filamentous bacteriophages (f1) to allow mutagenesis of the DNA in this system.

5.3 Cloning of viral genes

The basic principles of cloning and of expression of viral genes are illustrated in *Figure 5.2*. Viral DNA is purified or PCR amplified from the viral source and cloned into a plasmid. Complementary DNA (cDNA) copies of viral mRNAs or of viral RNA genomes may be produced to allow cloning. Other possible routes for cloning in bacteria include bacteriophage genomes or vectors derived from them, along with hybrid plasmid/bacteriophage systems with features of both systems. Systems derived using the DNA packaging signals of bacteriophage lambda (cosmids) allow the cloning of up to 45 kb of DNA, larger than almost all complete viral genomes.

After cloning into bacteria, clones containing the gene of interest are detected in purified plasmid DNA, usually by a combination of restriction enzyme mapping and the specific binding ('hybridization') of a DNA probe containing a base sequence complementary to a sequence present in the target DNA. A more thorough explanation of the methods used to clone, select and detect specific DNA is given in another book in this *Medical Perspectives* series: *Genetic Engineering*, and readers wishing to pursue this area further are referred to that book.

Following isolation of a clone containing the desired DNA, the gene of interest can be expressed in prokaryotic cells, using a plasmid vector, or transferred to a eukaryotic expression system (see Section 5.4). While expression in prokaryotes is more straightforward and is easier to scale up to commercial levels, there are many differences between prokaryotic bacterial cells and the far more complex eukaryotic cells. While proteins produced in prokaryotic cells will have the 'correct' amino acid sequence, it is important to remember that the amino acid chain (polypeptide) alone does not constitute a protein. Both during and after translation the amino acid chain undergoes a very wide range of other chemical modifications,

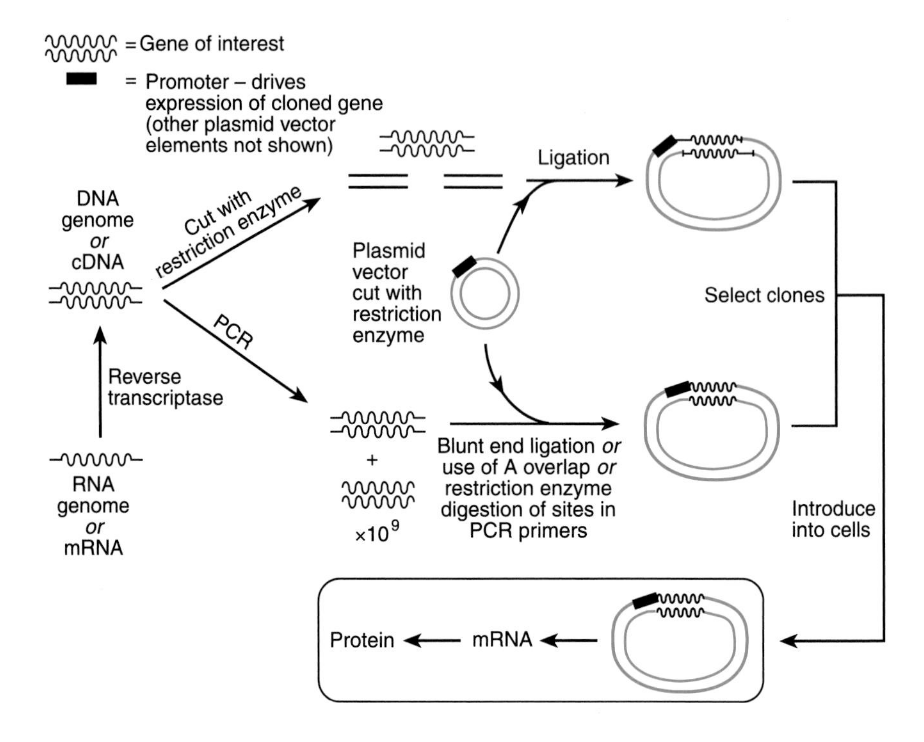

Figure 5.2: Cloning and expression of viral genes.

ranging from the attachment of small chemical groups such as phosphate or sulfate up to the attachment of complex sugar chains which can be bigger than the polypeptide itself. In addition, the environment of the protein translated may be important for correct folding. The amount and type of post-translational processing of a protein is determined by signals in the amino acid chain, but also varies with the type of cell, with cells from more complex organisms typically producing more complex modifications (see *Tables 5.1* and *5.2*). These modifications can have major effects on the structure and function of the protein. Folding of the

Table 5.1: Protein production systems

System	Authenticity[a]	Effort required	Scale-up potential	Maximum level of expression[b] (% of cell protein)
Bacteria	±	+	+ + +	30
Yeast	+ to + + +	+ +	+ + +	1–5
Baculovirus	+ +	+ + +	+	10 (30[c])
Mammalian cells	+ + +	+ + +	+	<1 (10[c])

[a]Similarity to viral proteins produced *in vivo*.
[b]As percentage of total cell protein production.
[c]In specific applications.

Table 5.2: Post-translational processing in protein production systems

	Bacteria	Yeast	Baculovirus/ insect cells	Mammalian cells
Precipitation	±	–	±	–
Protein folding	+	±/+	++	+++
Proteolysis	±	±/+	++	+++
Fatty acylation	–	++	++	+++
Phosphorylation	–	++	++	+++
Glycosylation	–	++[a]	++[a]	+++
Secretion	±[b]	++	++	+++
Function	±	++	++	+++

[a]Differs from that of mammalian cells.
[b]Signals differ from those used in eukaryotic cells.

protein can also vary, as illustrated by findings concerning the PrP^{Sc} protein of scrapie (see Section 1.8.6). Post-translational processing is very different in prokaryotic (as compared to eukaryotic) cells, with the result that the proteins produced in prokaryotic systems can be very different from 'authentic' proteins as produced when viruses infect eukaryotic cells (see *Tables 5.1* and *5.2*). These differences can result in poor solubility and in reduction or loss of immunogenicity or function. Where it is important to obtain protein with properties similar to those of the virus itself (for example in vaccine formulation), it is often necessary to express the gene of interest in a eukaryotic system. However, many functional proteins can be produced from prokaryotic expression systems, and these are extremely useful for expressing genes from viruses infecting eukaryotes.

5.4 Expression in eukaryotic cells

Expression of the gene of interest in eukaryotic cells will often involve subcloning, where the gene of interest is transferred to another plasmid which is capable of replication and expression in the eukaryotic system used. Alternatively, a plasmid may be used which contains a combination of controlling elements which allow it to replicate and express inserted genes in both prokaryotic and eukaryotic systems. Such a plasmid is sometimes referred to as a 'shuttle vector' and contains the necessary elements for both systems, as shown in *Figure 5.3*.

Like a prokaryotic vector, a typical plasmid for eukaryotic expression will have a promoter, a cloning site, a selectable gene and an origin of replication. Since many viruses are required to produce large quantities of their proteins in a relatively short time, many viral genes are expressed at very high levels. In addition, the factors controlling virus gene expression have been widely studied. As a result, many of the controlling elements in eukaryotic expression vectors are derived from viruses. Sequences from SV40 (polyomavirus) are widely used. Promoters can be high efficiency, and ensure a high level of transcription independent of the cell type, or can be promoters which are active only in certain cells or under specific

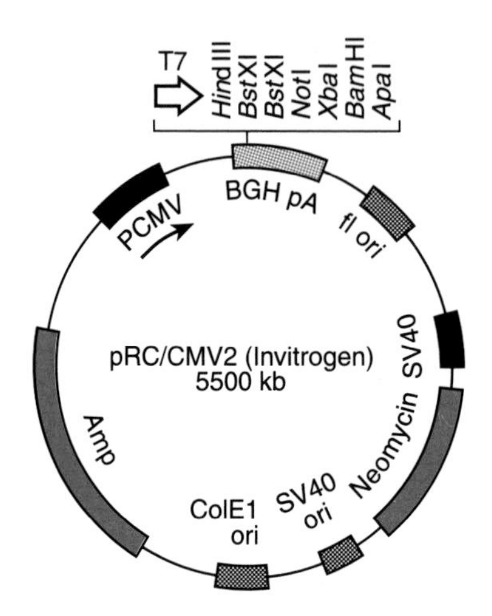

Figure 5.3: A prokaryotic/eukaryotic cloning vector. The stable eukaryotic expression vector pRC/CMV2 (Invitrogen) which can replicate in prokaryotic systems, showing:

Prokaryotic origin of replication (Col E1 ori)
Filamentous bacteriophage origin of replication (fl ori, for mutagenesis work)
Prokaryotic promoter (T7)
Prokaryotic selectable gene (amp)
Eukaryotic origin of replication (SV40 ori)
Eukaryotic promoters (CMV, SV40)
Eukaryotic selectable gene (neomycin)
Eukaryotic polyadenylation site (BGH pA)
Polycloning site (└┬┘) with the restriction enzymes cutting at that site listed

Note that this plasmid does not contain a reporter gene.

conditions, allowing expression of the inserted gene to be controlled. Examples of efficient promoters include the CMV immediate–early promoter or the Rous sarcoma virus LTR promoter, both of which are highly active in many cell types. An example of a conditional promoter is the metallothionein promoter, which is induced by heavy metals such as cadmium. Other eukaryotic elements equivalent to those seen in prokaryotic plasmid vectors are also present, including an origin of replication, cloning site and selectable gene. The origin of replication is also often of viral origin, while the selectable gene will differ from that seen in prokaryotic systems, but operates on similar principles. The most common eukaryotic selectable gene is one which gives resistance to neomycin (or the derivative, G418), an inhibitor of protein synthesis. One relatively new option is the use of a gene giving resistance to zeocin, an antibiotic that cleaves DNA in both prokaryotic and eukaryotic cells and is thus a selectable gene in both systems. As with prokaryotic vectors, the cloning site is a section of synthetic DNA containing multiple restriction sites. Additional features are also necessary for efficient expression in eukaryotic compared with prokaryotic systems. These include a polyadenylation site (often from a bovine growth hormone gene) adjacent to and downstream from the inserted gene which is required to produce functional eukaryotic mRNA, and which may also control splicing of the mRNA produced. Other regions of DNA which bind specific eukaryotic transcription factors and transcriptional activators are typically included to enhance transcription. An increasingly popular addition is the inclusion of a synthetic DNA sequence coding for six consecutive histidine residues in frame with the protein expressed and a proteinase site to allow their removal. This sequence binds the recombinant protein to a nickel affinity column, allowing rapid purification, followed by proteolytic removal of the histidine tag.

Unlike prokaryotic genes, eukaryotic genes are usually broken up by 'introns', regions of untranslated DNA that are removed from the pre-mRNA by splicing prior to translation. Viral genes generally have far fewer true introns than cellular genes, probably due both to limitations of space in the viral genome and to the involvement of viral factors which exploit this difference to enhance viral mRNA translation over that of cellular transcripts. Some viral vectors (see below) may not be able to express intron-containing genes (see *Table 5.3*). However, spliced viral mRNAs are not uncommon, and specific sequences, *splicing signals*, controlling splicing may be required.

The requirement for splicing of genes containing introns can be avoided by cloning a DNA copy (a 'cDNA') of an mRNA, which will already have had any introns removed. However, the lack of introns in such a clone may result in altered transport within the cell and inhibition of translation.

Table 5.3: Examples of widely used viral vector systems

Bacteriophage lambda	
Host cells	Prokaryotic (*Escherichia coli*)
Uses	Cloning, prokaryotic expression, construction of genomic libraries for screening
Advantages	Very wide range of vector systems available, very well characterized system, large inserts possible (up to 24 kb; up to 45 kb in the lambda-derived cosmid system)
Disadvantages	Size of recombinant DNA must be close to that of normal viral DNA, many eukaryotic events not possible (e.g. splicing out of introns, most post-translational processing events)
Bacteriophage M13	
Host cells	Prokaryotic (*Escherichia coli* expressing 'male' sex pili)
Uses	Based on single-stranded nature of M13 genome: sequencing, site-directed mutagenesis, DNA probe generation
Advantages	Single-stranded genome easily purified from extracellular medium, wide range of vector systems, intracellular genome acts as a plasmid, well characterized system, insert size usually 1–3 kb although larger in some systems
Disadvantages	Partial deletion or rearrangement of cloned DNA can occur with large inserts, one orientation of DNA may be preferred, no splicing of DNA, restricted insert size in most systems

(continued over page)

Table 5.3: *Continued*

Adenovirus	
Host cells	Human
Uses	Oral vaccine and gene delivery vectors, high efficiency protein production, production of transformed cell lines, infecting cells for immunological assay
Advantages	Oral delivery possible, authentic post-translational processing, replication-defective vectors (expressing only the insert) available, insert size typically up to 8 kb although 30 kb possible in some systems
Disadvantages	Not widely available, immunity to adenovirus prevents use, often cytotoxic
Baculovirus	
Host cells	Insect
Uses	High efficiency protein production
Advantages	Relatively easy to handle cells and virus, occluded virus can be grown in caterpillars, realistic post-translational processing (with some exceptions), large inserts possible (up to 15 kb), wide range of systems available, high protein expression
Disadvantages	Fusion proteins preferred, some post-translational processing may vary (notably glycosylation, although this is less pronounced than was suggested by early data), proteins produced at high levels may be insoluble, splicing may differ (introns should be avoided)
Herpesvirus	
Host cells	Very wide range of vertebrate cells
Uses	Neuronal targeting, vaccine vector, high efficiency protein production, episomal expression vectors (EBV-based)
Advantages	Well characterized viruses, larger inserts possible (up to 10 kb, larger in episomal vectors), wide choice of inserts and marker genes, authentic post-translational processing
Disadvantages	Biosafety concerns (virus latency), not widely available, highly cytotoxic, alter cell metabolism, transformation may occur
Retrovirus	
Host cells	Wide range of mammalian cells
Uses	High efficiency gene transfer, insertion of genes into cellular chromosome (gene therapy)
Advantages	Wide range of vectors available, very efficient introduction of DNA into cell, wide choice of promoters and marker genes, authentic post-translational processing
Disdvantages	Construction of vector may be complex, inserted genes must not interfere with retroviral replication and assembly, restricted insert size (less than 8 kb), concern over oncogenic potential and biosafety, requires replicating cells
Vaccinia	
Host cells	Human, primate, mammalian
Uses	Infecting cells for immunological assay, vaccine vector, expression of 'authentic' proteins
Advantages	Wide host range, authentic post-translational processing, choice of selectable markers, large inserts possible (up to 25 kb), high level expression is possible in hybrid systems, wide range of systems available
Disadvantages	Oligothymidine regions can cause early termination of transcription, introns not acceptable, possible biological hazard from vector virus, complexity of virus means that virus pathways (which may differ) are used instead of cellular systems, lack of nuclear involvement may be problematic for some proteins

5.5 Viral vector systems

As well as the use of viral sequences as components of plasmids, recombinant viruses can be constructed. These 'viral vectors' can carry foreign DNA inserted into their genomes, replicate (possibly with the aid of a helper virus or a cell line expressing viral genes), assemble and infect new cells. They may be constructed in both prokaryotic and eukaryotic systems. Viral vectors provide a ready-made, efficient system for getting the inserted DNA into cells without using the sometimes complex and inefficient chemical systems required with purified DNA. However, the effects of viruses on the host cell are often a problem, particularly virus-mediated inhibition of cellular synthesis and killing of infected cells, both by the virus and by cellular apoptosis (see Section 2.4) in response to infection.

Examples of widely used eukaryotic virus expression systems are shown in *Table 5.3*, with two prokaryotic systems also shown for comparison. As can be seen, different systems are suited to different applications. For example, for production of high levels of proteins in a eukaryotic system, placing an inserted gene under the control of the baculovirus polyhedrin promoter can result in the inserted gene accounting for up to 30% of cellular protein production in some cases. Baculoviruses infect insects, and in those used as expression vectors the polyhedrin gene is hyperexpressed late in infection to produce enough polyhedrin protein to form a protective capsule around the virus. By placing a gene under the control of the polyhedrin promoter, hyper-expression of the inserted gene can be achieved.

5.6 Gene therapy

Gene therapy involves the introduction of heterologous genes into the cells of an organism, usually to correct defective or non-operational genes within those cells. These may be the cells of an individual with an inherited genetic disorder, such as cystic fibrosis or hemophilia, or cells infected with or susceptible to a pathogen such as HIV, which are being altered to increase their resistance to the pathogen. There are several approaches to gene therapy, illustrated in *Figure 5.4*. Clearly an essential requirement for any of them is to have an identified and available therapeutic gene (often a functional version of a defective cellular gene), along with some way to deliver it to the cells where it is required.

The potential for gene therapy arising from the future results from the human genome project combined with ongoing developments in delivery systems will be immense. For use against pathogens, a wide range of therapeutic genes can be devised which inhibit replication or pathogenesis. These approaches overlap with some nucleic acid-based antiviral methodologies (see Section 4.7).

Delivery systems must be able to introduce DNA into the target cells. Where these cells are removed from the patient for treatment (exogenous gene therapy), or are within the patient (endogenous gene therapy) but in a readily accessible location, such as the lung cells of cystic fibrosis sufferers, chemical/mechanical systems of transfection, such as fusogenic

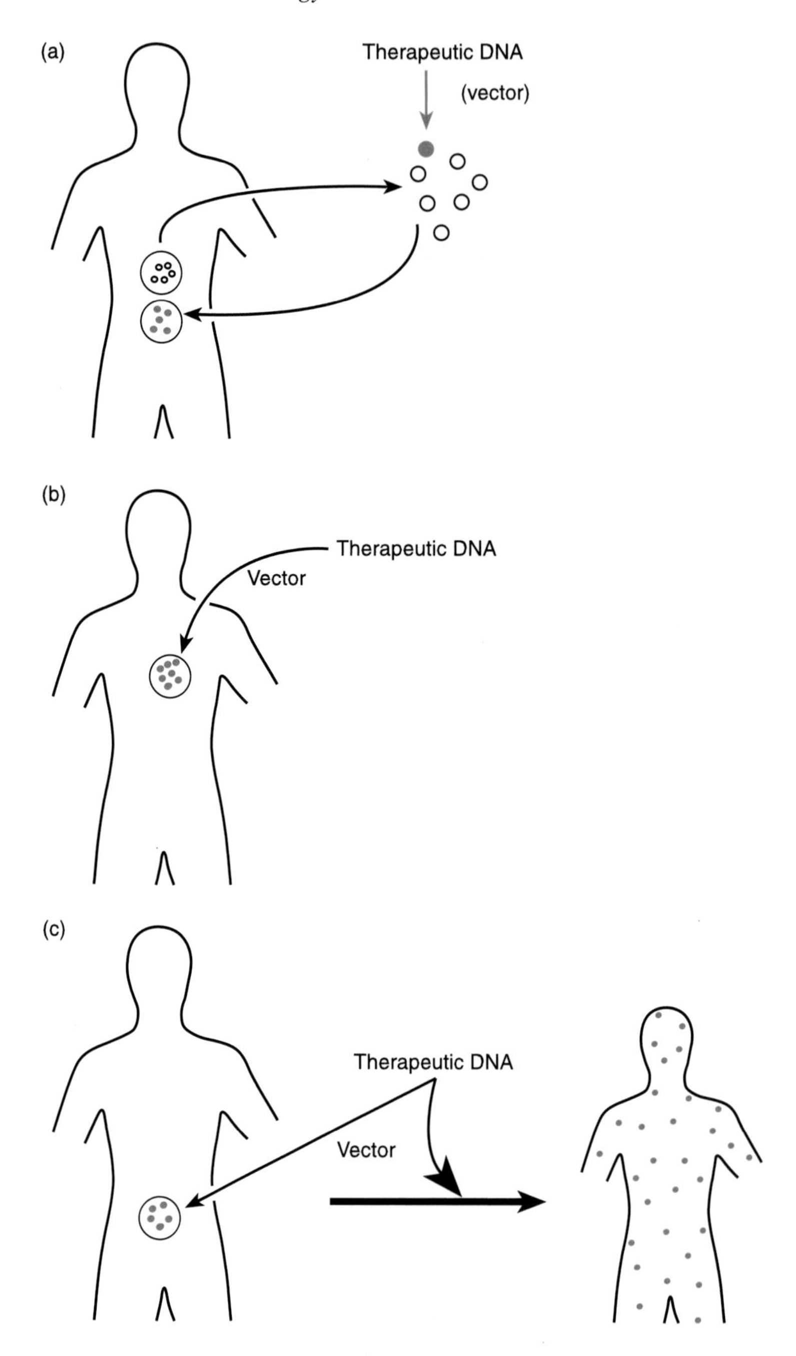

Figure 5.4: Approaches to gene therapy. (a) Exogenous somatic cell gene therapy, (b) endogenous somatic cell gene therapy, (c) germline gene therapy.

liposomes containing the therapeutic DNA, may be adequate. Once the DNA is within the cell it may integrate into the cellular genome, although this occurs at very low efficiency. Much early work used such systems on cells *in vitro*. However, as noted in Section 5.1, viruses provide highly efficient systems for getting foreign nucleic acid into cells. Also, viruses naturally exhibit cell tropism, where the requirements of the virus for specific receptors (see Section 1.4.1) along with other factors (such as the ability of the virus physically to get into certain areas of the body) can be used to ensure that specific types of cells are infected. While most viruses used for gene delivery will infect many types of cell, it is thought that careful selection (and, where appropriate, modification) of the virus vector will ensure that the therapeutic gene is delivered to a specific location. One example of such a delivery system is the targeting of neuronal cells using vectors based on HSV, but this has not as yet lived up to its original promise. As yet, adenoviruses appear to provide the most practical virus vector system, and while adenoviruses themselves show little specificity in the cell types that they infect, work is under way to engineer such specificity into the system, either by altering the surface proteins of the virus or by using cell type-specific promoters to control the inserted gene. Once the therapeutic gene is inside the target cell, it must be expressed at an appropriate level. Once again, viruses can provide a route to achieve this. Many viruses, such as retroviruses or the adeno-associated (parvo) virus, have a high efficiency integration step in their life cycle, and foreign genes introduced into the viral nucleic acid may use viral mechanisms to become integrated into the cellular DNA. Other viruses (herpesviruses, papillomaviruses) are maintained stably as extrachromosomal genetic elements (episomes). Therapeutic genes present in integrated or episomal DNA can be stably expressed in cells. Virus vectors have been used for exogenous gene therapy, but it is where the target cells must be treated in less accessible areas of the body that virus vector systems are likely to be of most value. Viruses being evaluated for use as gene therapy vectors are summarized in *Table 5.4*. Despite their possible oncogenic potential, retrovirus systems show a great deal of promise since they are very efficient integrating vectors (relative to the usual level of DNA integration) with a wide range of available cell tropisms.

Current approaches are based around somatic cell therapy, where cells from the site where the genetic defect is pathogenic are treated with therapeutic DNA. Examples include the lung cells of cystic fibrosis sufferers where a defect in an ion transport protein damages the lung surface, and the hematopoietic stem cells of sufferers from adenosine deaminase deficiency, which causes profound immunosuppression by killing lymphocytes. Specialization of cells (differentiation) may result in variant gene expression, complicating attempts at gene therapy. For example, lung cells are highly specialized. It is also important to note that the routine elimination of cells by the body (the rate of which varies enormously between cell types) can require relatively frequent repetition of gene therapy since cells expressing the therapeutic protein are not immune to routine replacement by new, untreated cells. Such replacement is relatively frequent (but appears not to be too high to permit therapy) in the case of

Table 5.4: Possible viral vectors for gene therapy

Retroviruses
Introduce DNA into the cell very efficiently
Integrate at multiple sites in the cellular genome (some may be more restricted)
May be oncogenic
Restricted insert size
Require dividing cells for transduction
Parvoviruses (adeno-associated virus)
Integrate at a few sites (usually only one) in the cellular genome
Non-pathogenic in humans
Very restricted insert size
High titers of virus may be obtained
Adenoviruses
DNA may integrate
Restricted insert size
Possibly oncogenic
Adenovirus-specific immunity may destroy treated cells
Large inserts possible in some systems
Herpesviruses
Maintained as episomes in latently infected cells
Can be targeted to neural cells, but may be neuropathogenic
May be oncogenic
Highly cytopathic
Very large inserts possible
Papillomaviruses
Maintained as episomes
Restricted insert size
Can be targeted to neural cells
May be oncogenic

lung cells, but the extremely high rate of turnover of lymphocytes would make them less suitable targets. By delivering the therapeutic gene into the hematopoietic stem cells which actually produce all of the cell types in the blood, this problem is avoided while the number of cells produced which contain the therapeutic gene is greatly increased. Unsurprisingly, hematopoietic stem cells represent a major target of gene therapy.

This leads logically into the ultimate gene therapy. There is only one precursor cell type that will ensure that *all* cells produced contain the therapeutic gene: the germ cells. Clearly even the possibility of establishing foreign DNA in the human germline, and (at least potentially) in every copy of the human genome descended from the treated individual, is something that requires thorough knowledge along with extremely careful consideration. While the technology has been established and transgenic animals produced, there are as yet no serious proposals for human germline therapy.

As will be appreciated, any work with recombinant DNA requires careful assessment of the risks and benefits and of the ethical issues involved. Codes of conduct and supervisory bodies have been established in many countries. Gene therapy, where the intention is to introduce and express recombinant DNA in humans, is one of the most controversial areas.

Despite this, by 1996, 163 protocols had been approved in the United States and Europe, of which 119 were aimed at cancer treatment, using a wide range of methodologies. In theory, there are few biomedical problems which could not be addressed by gene therapy. It is the promise of successful gene therapy which underlies much of the current work on molecular virology.

Further reading

Davison, A.J. and Elliott, R.M. (1993) *Molecular Virology: a Practical Approach*. IRL Press, Oxford.

Kinchington, P.R. (1993) Expression of virus genes in heterologous systems. In *Virology Labfax* (D.R. Harper, Ed.). BIOS Scientific Publishers, Oxford, pp. 265–286.

Lemoine, N.R. and Cooper, D.N. (1996) *Gene Therapy*. BIOS Scientific Publishers, Oxford.

Patterson, R.M., Selkirk, J.K. and Merrick, B.A. (1995) Baculovirus and insect cell gene expression: review of baculovirus biotechnology. *Environ. Health Perspect.*, **103**, 756–759.

Sambrook, J., Fritsch, E.F. and Maniatis, T. (1989) *Molecular Cloning: A Laboratory Manual*. Cold Spring Harbor Laboratory Press, Cold Spring Harbor, NY.

Williams, J., Ceccarelli, A. and Spurr, N. (1993) *Genetic Engineering*. BIOS Scientific Publishers, Oxford.

Electronic resources

For specific sources, use of a search engine such as Alta Vista or Webcrawler will provide links to relevant sites and, since URL addresses change frequently, may be more up to date than the links provided below. Some useful examples are listed below, and provide a variety of links to related material:

The EMBL Nucleotide Sequence Database
http://mercury.ebi.ac.uk/ebi_docs/embl_db/ebi/topembl.html

National Center for Biotechnology Information, Genbank overview
http://www.ncbi.nlm.nih.gov/Web/Genbank/

PIR International Protein Sequence Database
http://www.bis.med.jhmi.edu/Dan/proteins/pir.html

SEQNET, SEQuence NET work computer for molecular biologists
http://www.seqnet.dl.ac.uk/home.html

Swiss-Prot, annotated protein sequence database
http://expasy.hcuge.ch/sprot/sprot-top.html

Chapter 6

Molecular diagnostics

Many of the techniques used for routine viral diagnosis and shown in *Table 6.1* fall outside the scope of this book. Owing to limitations of space, only those based on molecular techniques (in particular nucleic acid detection or amplification) will be covered in any detail in this chapter. Interested readers are referred to the Further Reading section at the end of the chapter for details of diagnostic techniques not covered here.

6.1 Immunological assays

Immunoassays are one of the oldest techniques of diagnostic virology, with detection of antibodies in patient sera being indicative of past (IgG) or recent (IgM) infection with the agent in question. Viral antigens may also be detected by immunological tests. However, the development of monoclonal antibodies has revolutionized this area, and a list of the diagnostic kits now available which use monoclonal antibodies (see Section 2.8.2) would be far too long to be included in a book of this size. Additionally, the sensitivity and specificity of techniques such as immune electron microscopy, immunofluorescence and immunocytochemistry are greatly enhanced by the use of monoclonal antibodies.

The production of test antigens for diagnostics has also benefitted from molecular techniques. Purification of antigens produced in tissue culture can be difficult, and many viruses are troublesome or impossible to grow in culture, making it very difficult to obtain antigens. Cloning and expression of viral genes provides a relatively straightforward alternative approach, simplifying purification and large-scale production. Indeed, as noted in Section 6.3 for hepatitis C, a previously unknown agent can be identified and a specific assay developed entirely by molecular techniques. As with monoclonal antibodies, it would be difficult (and rapidly out of date) to list all of the kits based on the use of recombinant antigens. The systems used to prepare and evaluate such antigens are substantially as those for vaccine components (see Chapters 3 and 5), and indeed the same antigens may find applications in both systems. Many of the problems

Table 6.1: Commonly used diagnostic techniques

Technique	Specimen types	Advantages	Disadvantages
Antibody titration (ELISA, RIA, IF, CFT Western blot, many other techniques)	Serum, CSF	Rapid; applicable to almost all viruses; inexpensive; can detect prior infection or demonstrate markers of immunity; detection of IgM shows recent infection	Identifies immune response to virus rather than virus itself
Virus culture	Blood, (buffy coat), CSF, stool, vesicle fluid, BAL, NPA throat washings, urine, swabs	Detects multiple viruses, but some will not grow in culture (e.g. papillomaviruses, hepadnaviruses); 'gold standard' for virus detection; only detects infectious virus	Slow; labor intensive; expensive; specimen transport conditions can be critical
Accelerated culture (virus centrifuged on to cell sheet)	As virus culture	As virus culture, but faster	Very labor intensive
Culture plus immunodetection	As virus culture; requires specific antisera	As virus culture, but much faster	Requires specific antisera; very labor intensive
Electron microscopy	Vesicle fluid, stool, cultured virus, skin scrapings, (urine)	Rapid	Only usable where virus is present at very high concentrations; cannot discriminate between morphologically similar viruses; high capital and staff costs
Electron microscopy plus immunodetection	As electron microscopy	Rapid; can discriminate between similar viruses	Requires high concentrations of virus and specific antisera; labor intensive; expensive
Immuno-cytochemistry	Biopsy specimens, vesicle fluid, skin scrapings	Rapid; shows infected cell type; applicable to most viruses	Requires specific antisera
Nucleic acid detection (ISH, electrophoresis)	Biopsy specimens, stool	Rapid; can show infected cell type; applicable to most viruses	Requires specific (labeled) nucleic acid probe

Table 6.1: *Continued*

Technique	Specimen types	Advantages	Disadvantages
Nucleic acid amplification (PCR, NASBA)	Any specimen	Rapid; extremely sensitive; applicable to most viruses	Requires knowledge of viral genome sequence; requires good controls to avoid false positive results; can detect insignificantly low levels of virus; labor intensive; viruses with RNA genomes require reverse transcriptase step before PCR

BAL, bronchoalveolar lavage; CFT, complement fixation test; CSF, cerebrospinal fluid; ELISA, enzyme-linked immunosorbent assay; IF, immunofluorescence; ISH, *in situ* hybridization; NASBA, nucleic acid sequence-based amplification; NPA, nasopharyngeal aspirate; PCR, polymerase chain reaction; RIA, radioimmunoassay.

associated with the use of recombinant antigens are again similar to those associated with vaccine development, including the production of 'incorrectly' processed antigens by most expression systems and the problem of using a single cloned antigen to detect a viral antigen that may vary between isolates.

6.2 Nucleic acid detection and amplification

Nucleic acid and antigen detection systems are fundamentally different from antibody assays, since they detect a component of the organism itself, rather than serological evidence of its past presence. However, viral genetic material may still be present even in the absence of infectious virus. With some viruses, it is possible to detect low levels of the viral nucleic acid long after the initial infection, while others are cleared rapidly from the body during the resolution of the infection. There are several techniques used to detect viral nucleic acids. These include *in situ* hybridization, dot-blot hybridization and the polymerase chain reaction (PCR). In addition, the techniques of gel electrophoresis and blotting may be used for specific applications.

The nature of the viral genetic material must be considered, since many viruses have RNA genomes, and do not have any DNA stages in their life cycle, necessitating the detection of RNA. With cellular organisms, RNA detection is not commonly used for diagnostic purposes, but is rather a research tool used to determine levels of mRNA and transcriptional activity. However, with many viruses, RNA detection is required.

6.2.1 Direct detection of nucleic acids

Direct detection of viral nucleic acids from clinical specimens is shown in *Figure 6.1*. Such techniques are rarely used at present, although epidemiological typing of dsRNA genome segments purified from rotaviruses (from stool specimens) is possible. Rather more common is the assay of restriction fragment length polymorphisms (RFLPs) in the

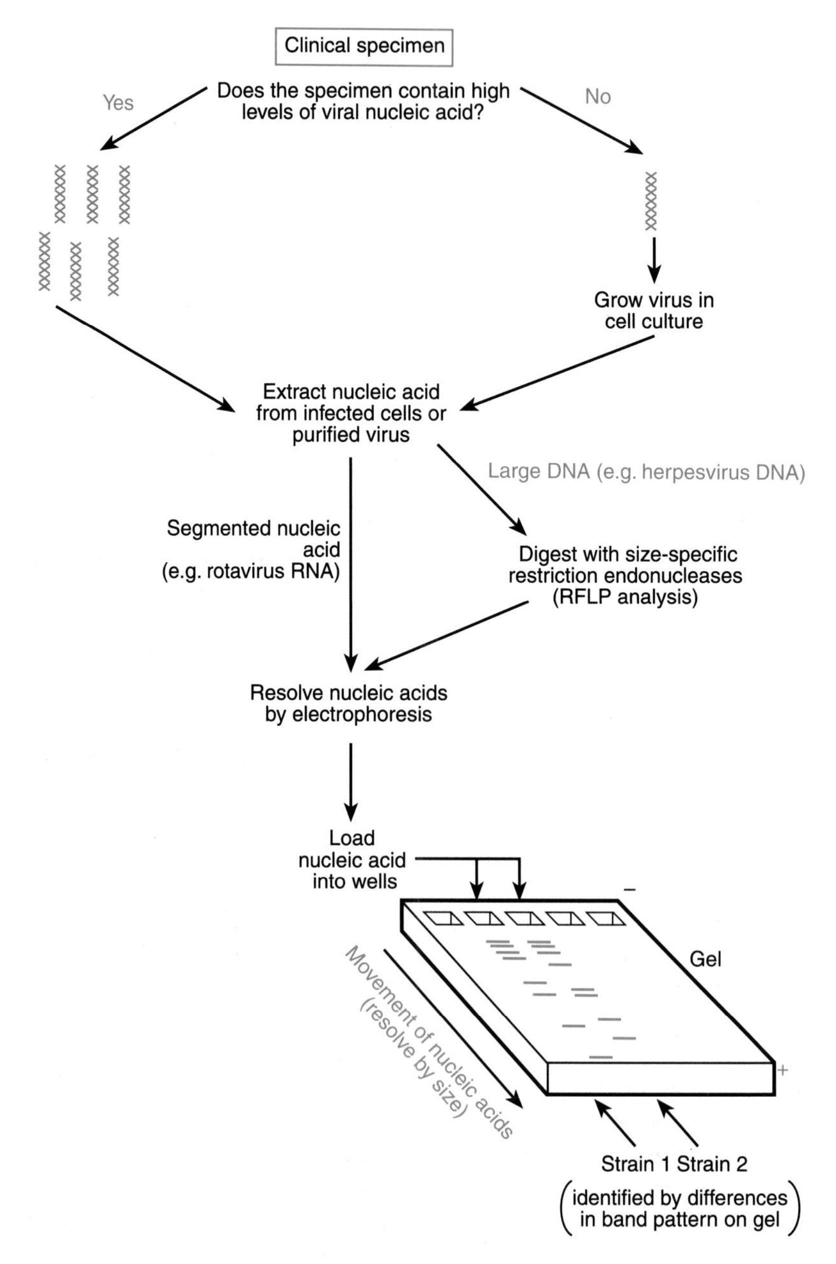

Figure 6.1: Direct detection of nucleic acids.

genomes of cultured virus. This technique is also shown in *Figure 6.1* and is only applicable to viruses with DNA genomes, although DNA copies (cDNAs) of RNA genomes can be analyzed. Restriction endonuclease enzymes are used to cut the DNA at specific sites. Changes in DNA sequence at a cut site will produce changes in the number and size of bands produced. This method is also referred to as restriction endonuclease analysis (REA). RFLP analysis has been used to type strains of many viruses, and may represent the most convenient means of discriminating between strains of some viruses. However, diagnostic assay of clinical specimens usually requires the additional sensitivity of hybridization- or amplification-based systems.

6.2.2 Hybridization

Hybridization uses a labeled 'probe' nucleic acid with a sequence complementary to that of the 'target' nucleic acid to be detected. This means that the bases of the probe will pair with those in the target nucleic acid, if it is present. The bound probe is detected as described below. For dot-blot or *in situ* hybridization, nucleic acid probes may be purified from a biological system containing the desired sequence, usually in the form of a plasmid containing cloned viral genes. Alternatively, short probe oligonucleotides (typically up to 35 nucleotides in length) may be synthesized directly from nucleotide monomers.

In order to be detected, probes must be labeled in some distinctive way. For nucleic acid probes, this was traditionally the incorporation into the probe during synthesis of radiolabeled nucleotides, most often using radioactive phosphorus (^{32}P). Subsequent detection by radiometric quantitation or exposure to X-ray film is fast and simple. However, there are a number of problems associated with radioactive detection systems. Handling of radioisotopes requires specialist facilities and safeguards, and many (notably ^{32}P) have inconveniently short half-lives. Recent developments have shown that non-radioactive detection systems can have comparable sensitivity, and the popularity of such techniques is increasing. Some systems use a small 'reporter' molecule bound to the probe, with a larger 'reporter binding' protein–enzyme conjugate added later. Other systems use a colored, fluorescent or enzymic label bound directly to the probe. If an enzyme label is employed, this is then used to produce a colored or luminescent reaction product, which can be detected visually or by an automated reading system. The range of possible probe labeling systems is too large to cover here, but in a diagnostic setting it is worth noting the current trend away from radioactive labels, with luminescence in particular becoming more popular.

The principal diagnostic nucleic acid hybridization systems are *in situ* hybridization and dot-blotting. In these techniques, a probe nucleic acid is used to assay for the presence of a matching DNA or RNA sequence in fixed cells (*in situ* hybridization) or in purified nucleic acid bound to a filter membrane (dot-blotting). Dot-blotting merely provides a quantitation of the amount of matching nucleic acid in the tissue sample from which it is extracted, while *in situ* hybridization also provides information on the cell type infected and on the intracellular location and quantity of viral nucleic acid present. Both systems can detect latent or inactive virus, and can be of comparable sensitivity to immunological detection systems, without the need to develop specific antibodies.

An enhanced form of hybridization detection, the branched DNA system, uses a probe that has multiple binding sites for a second, reporter probe, enhancing the resulting signal. This is broadly similar to the use of peroxidase–antiperoxidase complexes in protein detection.

6.2.3 Nucleic acid amplification

Amplification-based techniques differ from hybridization-based detection systems in that probes are modified (usually by extension) following binding to the target nucleic acid sequence. This process is then repeated in a series of cycles, typically allowing all nucleic acids produced in one reaction to function as templates for the next, giving a 'chain reaction'. Over time and with repeated cycles, this results in an exponential increase in the amount of target-complementary nucleic acid, with billion- to trillion-fold (10^9–10^{12}) amplification of this one specific sequence. As with hybridization-based systems, the specificity of the reaction comes from the selective nature of the binding of the short synthetic nucleic acid probes, referred to in this setting as primers, to the target nucleic acid. The large number of copies produced makes detection relatively simple.

The polymerase chain reaction. PCR was the first amplification-based system to be developed, and was developed by Dr Kary Mullis in 1985 while working for the Cetus Corporation in California. As a result, the basic technique of PCR has, unlike most diagnostic techniques, been protected by patent almost since its inception. These patents are now held world-wide by Hoffman LaRoche, and use of PCR requires licensing by the company. The basic principles of PCR are summarized in *Figure 6.2* and *Table 6.2*, while some enhancements to the original technique are shown in *Table 6.3*.

In the earliest PCR work, the polymerase used was a fragment of the *Escherichia coli* DNA polymerase (the 'Klenow fragment'). This required the addition of fresh polymerase after every heating cycle. It was only with the use of heat-resistant DNA polymerase isolated from the thermophilic bacterium *Thermophilus aquaticus* that PCR became a practical technique, since one addition of the polymerase at the start of the reaction is all that is required. Since purifying the enzyme from the thermophilic bacterium was expensive, a cloned form of this '*Taq* polymerase' was developed, and is now more widely used. A range of other heat-resistant polymerases are also available, offering advantages over the *Taq* enzyme in thermostability, maximum product size, or fidelity of replication, but the nature

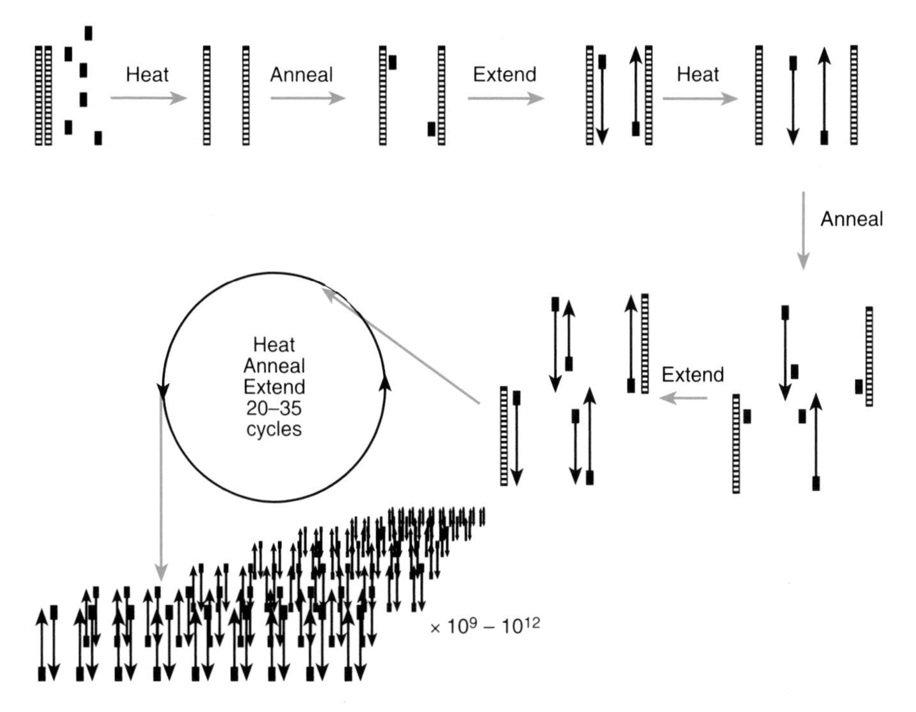

Figure 6.2: Polymerase chain reaction (PCR). (▮) Primer, (▤) target nucleic acid, (↑↓) PCR products.

Table 6.2: Polymerase chain reaction

The original system, designed for DNA detection. RNA can be detected if an initial RNA-to-DNA copying (reverse transcriptase) step is included or if an enzyme with innate reverse transcriptase activity is used
Reaction cycle is:
Heat – separates DNA strands
Anneal – two oligonucleotide primers bind to opposite strands of DNA, usually 100–500 bp apart
Extend – DNA extended across region between primers
Automated heating block (thermal cycler) used
Heating step requires the use of thermostable enzyme (see *Table 6.4*)
Detection by electrophoresis, hybridization or hybridization in microplate (may use capture tag on primer)
Kits available from Hoffman LaRoche for HIV-1, HTLV-1 and 2, hepatitis C virus, CMV, enteroviruses. Reverse transcriptase–PCR using *Tth* polymerase (see *Tables 6.3* and *6.4*) used for RNA detection
Kits available shortly for papillomaviruses and cytomegalovirus
Quantitative PCR (see *Table 6.3*) kits available for HIV-1 and hepatitis B and C viruses
Multiple enhancements available (see *Table 6.3*)

of the PCR patent casts some doubt on the use of such enzymes. Some of the enzymes available for PCR are summarized in *Table 6.4*.

Alternative amplification-based detection systems. Alternative systems have now been developed by other companies primarily for diagnostic uses,

Table 6.3: Enhancements of PCR

Reverse transcriptase (RT) PCR

Initial RT step copies RNA to DNA (*Tth* thermostable DNA polymerse has inherent RT activity)

Advantages

Allows detection of RNA (including viral RNA genomes) by PCR

Disadvantages

May not be as sensitive as direct DNA detection

Multiplex PCR

Uses multiple primer sets in one tube to allow simultaneous assay for multiple target nucleic acids

Advantages

Allows one test to detect multiple agents simultaneously
Primers used can be selected to detect a range of agents appropriate for particular specimen types or clinical conditions

Disadvantages

Each primer set must be highly specific
Requires thorough optimization of individual PCR assays
Primer set reaction conditions must be compatible
Non-specific bands can cause confusion

Nested PCR

Product of first PCR is used as template for a second PCR prior to detection

Advantages

Added sensitivity from double PCR
Added specificity from second primer binding stage

Disadvantages

Increased contamination from aerosols generated during second stage PCR set up (uracil *N*-glycosylase cannot be used at this stage since PCR product is being assayed for)
Increased cost

In situ PCR

Fixed, protease-digested cells are diffused with PCR reaction mixture, PCR product does not diffuse

Advantages

Location of template within cells is shown

Disadvantages

Complex sample preparation and reaction conditions
Increased cost

Quantitative PCR

Uses co-amplified template with small modification allowing discrimination of product from that amplified from natural template. Multiple reactions using different levels of co-amplified template are compared with amount of product from natural template to quantify amount present

Advantages

Quantitation shows amount of viral genome present ('viral load')

Disadvantages

Multiple reactions required
Modified template required for co-amplification
Increased cost

Table 6.4: Examples of thermostable polymerases used in the PCR

Polymerase (supplier)	Source	Thermostability (half-life (min) at 95°C)	Fidelity (proof-reading activity)	Comments
Taq (Cetus, others)	*Thermophilus aquaticus*	High (40)	Moderate (no)	The original thermostable polymerase
Amplitaq® (Cetus)	*Taq* gene expressed in *Escherichia coli*	High (40)	Moderate (no)	The most widely used and best characterized
Vent™ (New England Biolabs)	*Thermococcus litoralis* gene expressed in *E. coli*	Very high (400)	High (5–15 × *Taq*) (yes)	Suitable for producing larger fragments (8–13 kb)
Deep Vent™ (New England Biolabs)	*T. litoralis* gene expressed in *E. coli*	Extremely high (1380)	High (5–15 × *Taq*) (yes)	Suitable for producing larger fragments (8–13 kb)
Pfu (Stratagene)	*Pyrococcus furiosus*	Very high (> 120)	High (12 × *Taq*) (yes)	
Tth	*Thermus thermophilus* gene expressed in *E. coli*	Moderate (20)	Moderate (no)	Enzyme has inherent reverse transcriptase activity and is used for PCR of RNA
UlTma™	Modified *Thermotoga maritima* gene expressed in *E. coli*	High (> 50)	High (yes)	

Derived from Newton and Graham (1997) *PCR*, 2nd Edn, BIOS Scientific Publishers.

and are summarized in *Figures 6.3–6.5* and *Tables 6.5* and *6.6*. Nucleic acid sequence-based amplification (NASBA) allows direct detection of RNA without a PCR step and kits are available. Related systems are also under developent. No ligase chain reaction (LCR) kits are available for viruses, and other kits are available only for the LCx automated diagnostic system.

Detection of amplified nucleic acids. Originally, detection of amplified DNA with PCR was on the basis of size by electrophoresis. This is labor intensive

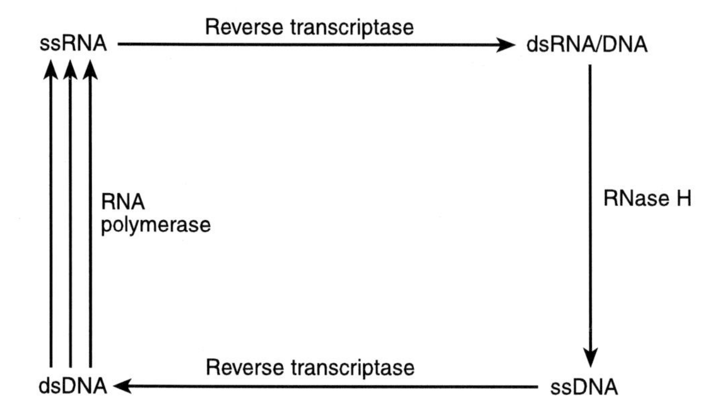

Figure 6.3: Self-sustained sequence replication (3SR) or nucleic acid sequence-based amplification (NASBA) (primers initiate both DNA and RNA synthesis).

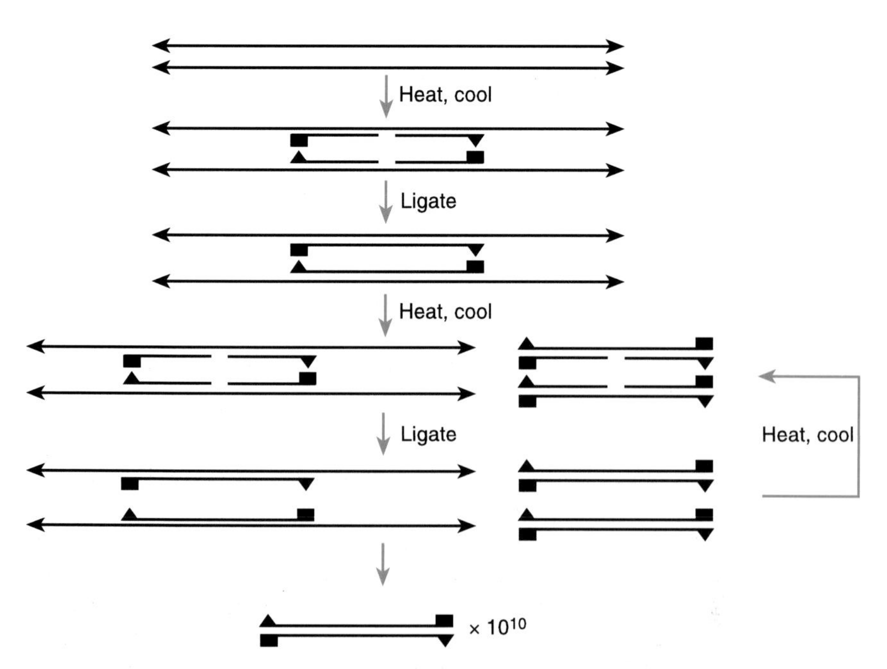

Figure 6.4: Ligase chain reaction (LCR). (▼) Capture tag, (■) reporter tag.

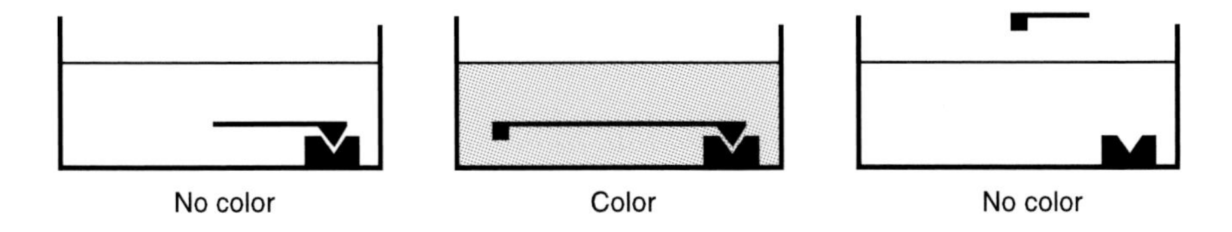

Figure 6.5: LCR detection (ELISA-type format). (▼) Capture tag, (■) reporter tag.

Table 6.5: Nucleic acid sequence-based amplification (NASBA)

Also known as self-sustaining sequence replication (3SR) Development of transcription amplification system (TAS)
Designed for RNA or ssDNA. Can work for dsDNA if an initial heating step is included
Reaction cycle is:
two oligonucleotide primers with terminal extension allowing T7 RNA polymerase to attach/bind to opposite strands of DNA 100–500 bp apart RNA copied to dsDNA/RNA hybrid by retroviral reverse transcriptase RNA degraded by bacterial ribonuclease active only on DNA/RNA hybrids (RNase H) ssDNA copied to dsDNA by reverse transcriptase dsDNA transcribed by bacteriophage (T7) RNA polymerase binding to primer extension
While reaction was designed to be isothermal at physiological temperatures, and not to require a thermal cycler, cyclical heating and cooling may increase sensitivity
Detection as PCR
Kit available from Organon Teknika for detection of HIV-1 and CMV (late mRNA) Quantitative NASBA kit for HIV-1 also available

Table 6.6: Ligase chain reaction (LCR)

Developed for microplate diagnostic systems
Reaction cycle is:
Heat – separates DNA strands Anneal – four oligonucleotide primers bind to adjacent regions of DNA, two on each strand, at matching locations Ligate – adjacent primers are joined by thermostable ligase
Thermal cycling requires the use of thermostable enzyme
Detection by hybridization in microplate using capture tag on one primer and reporter tag on ligated primer
Developed by Abbott Laboratories, no kits available for viruses, method used only in automated LCx system

and provides no information on the PCR product other than its size(s). As a result it was often combined with Southern blotting, where the resolved nucleic acids were transferred to a membrane and reacted with nucleic acid probes specific for the desired PCR product (*Figure 6.6*). This was very labor intensive and time-consuming, and not well suited to diagnostic use.

More recently, microplate systems based on specific hybridization have been developed which avoid the need for complex and time-consuming electrophoretic procedures (*Figure 6.7*), and these are more

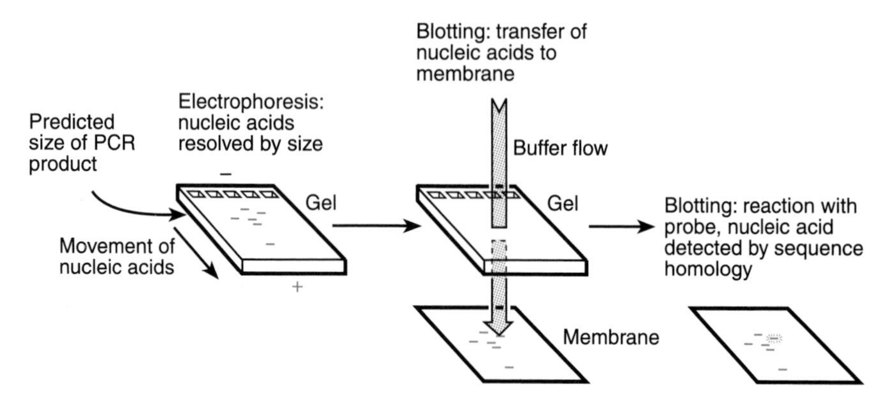

Figure 6.6: Electrophoresis and Southern blotting.

suited to diagnostic use. An alternative system is based on scintillation assay of radiolabeled PCR product, binding to beads containing a scintillant that emits light when struck by radioactive emissions (the scintillation proximity assay). Some of these systems also allow some quantitation of the amount of nucleic acid template present in the initial reaction mixture.

Problems of amplification-based nucleic acid detection systems. All amplification-based systems can provide exquisitely sensitive diagnostic systems which can detect latent or inactive virus. However, they may detect insignificant levels of virus, can only be used diagnostically to detect viruses of known nucleic acid sequence and are extremely sensitive to contamination, since extremely low levels of contaminating nucleic acid will give false-positive results. Methods of avoiding contamination are summarized in *Table 6.7*. The PCR is also able to generate nucleic acids for cloning rapidly and without any requirement for convenient restriction sites. Low polymerase fidelity may be a problem in this application, but higher fidelity polymerases, which 'proof-read' the newly synthesized DNA and correct any mistakes, are now available (see *Table 6.4*).

6.3 Future developments in diagnostic virology

It is very clear that nucleic acid detection systems, particularly those based on amplification, are likely to play an increasing part in viral diagnosis. However, antibody and antigen assays, increasingly based on the use of cloned proteins, will continue to play a prominent role.

There is an increasing trend away from 'in-house' assays, towards the use of commercially produced assays (kits). Although such kits are evaluated and certified by the producing company, their performance must be evaluated by the user. However, the manufacturer's quality control may be useful in obtaining laboratory accreditation. Kits may also be useful where facilities and expertise for assay development are not available, and they are better suited to use with automated systems.

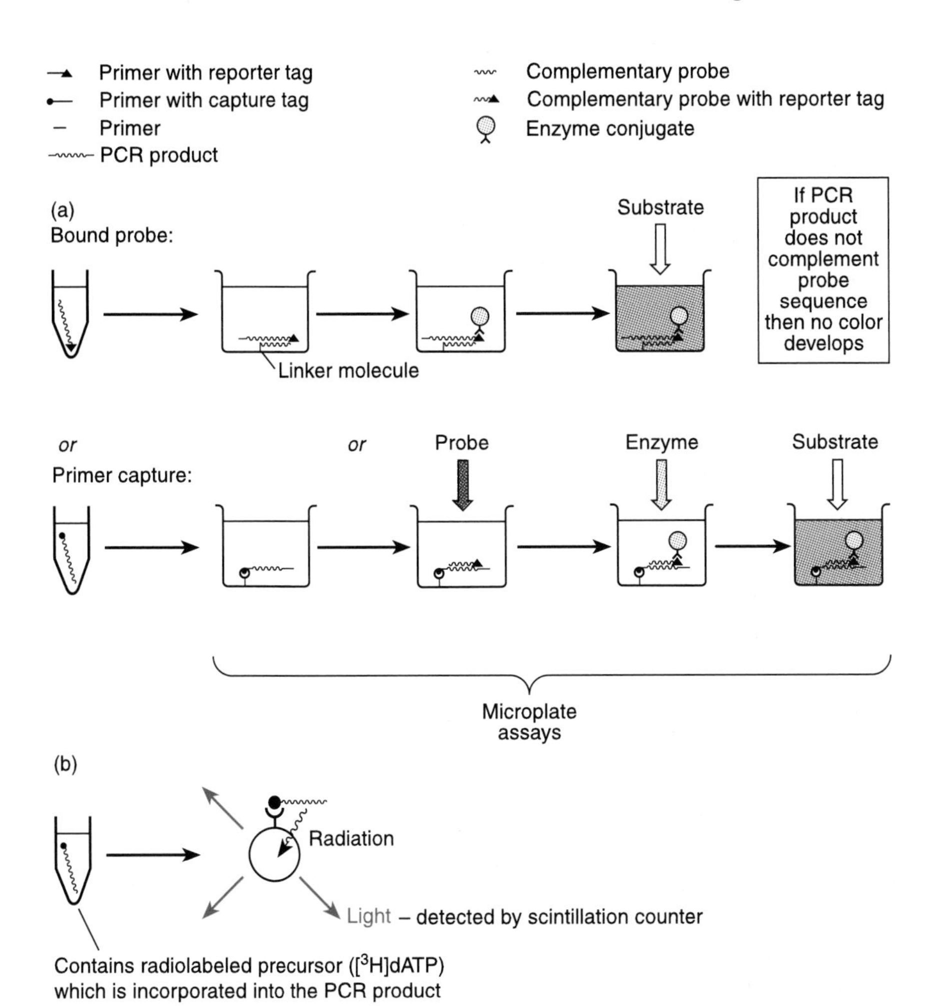

Figure 6.7: Non-electrophoretic detection systems for PCR product. (a) Microplate hybridization; (b) scintillation proximity assay (SPA).

Table 6.7: Countermeasures to contamination in amplification-based systems

- Inclusion of a full range of negative controls interspersed with test samples (positive controls must also be included, but should be carefully segregated from test samples)
- Separation of preparation, set up and analysis areas
- Separation of preparation, set up and analysis equipment
- UV irradiation of work areas, reaction mixtures (before nucleic acid addition) and equipment
- Use of fresh (autoclaved or sterilized) reagent aliquots, disposable containers, plugged or positive-displacement pipet tips
- UNG (uracil *N*-glycosylase) combined with uracil (rather than thymidine) in PCR reaction mixture: destroys contaminating PCR product
- Hot-start PCR: wax barrier prevents reaction start until high temperature is attained, minimizes non-specific priming
- Anti-taq blocking antibody: inactivates *Taq* until irreversibly removed by initial heating step

Automated analytical systems represent another strong trend in diagnostics. The newest machines are complex and flexible, and are capable of handling multiple assays simultaneously while performing all stages of testing from specimen preparation to issuing of test results. However, the tests that can be used with such a system are of course limited to those provided by the manufacturer, which can be restrictive, particularly where the laboratory has specialized requirements. In addition, the machines are extremely expensive, although this is moderated by potential savings in staff costs and by package deals linked to purchases of assay kits. Indeed, the latter may offset much of the initial cost, although long-term contracts may be required. Specimen turn-round times may be shorter than with manual testing, and 'out of working hours' testing may be simpler to arrange.

One interesting area for development is the application of sequence data to diagnostic virology. Clearly, knowledge of a viral genome sequence is required for most nucleic acid detection systems. Such knowledge has traditionally followed long after the virus itself was detected and characterized, but this is changing. In the case of hepatitis C virus, it was known that blood could transmit some forms of hepatitis not linked to either of the two hepatitis viruses then known (hepatitis A and hepatitis B). A 'virus fraction' was prepared from plasma containing this 'non-A non-B' (NANB) hepatitis, concentrated, and extracted nucleic acids were reverse transcribed, cloned and expressed. The resulting protein product was reacted with sera from patients, and specific antibody was found to be present in a high percentage of those who had suffered from transfusion-associated NANB hepatitis. A diagnostic kit was developed (and patented) around this protein (expressed in yeast cells), and it was only after this body of work that anyone actually saw the virus itself. It has been classified as a member of the *Flaviviridae* (see Section 1.2) and is clearly a major agent of human disease. Other molecular techniques have already identified human herpesvirus 8, proposed as the causative agent for Kaposi's sarcoma (see Section 1.7), emphasizing that there are alternative approaches to the 'classical' pathway of isolating and propagating an infectious agent in culture. Molecular virology has clearly demonstrated its clinical value.

Further reading

Barany, F. (1991) The ligase chain reaction in a PCR world. *PCR Methods Appl.*, **1,** 5–16.

Compton, J. (1991) Nucleic acid sequence-based amplification. *Nature,* **350,** 91–92.

Mullis, K.B. (1990) The unusual origin of the polymerase chain reaction. *Sci. Am.*, **262,** 56–65.

Newton, C.R. and Graham, A. (1997) *PCR*, 2nd Edn. BIOS Scientific Publishers, Oxford.

Schmidt, N.J. and Emmons, R.W. (Eds) (1989) *Diagnostic Procedures for Viral, Rickettsial and Chlamydial Infections.* American Public Health Association, Washington, DC.

Electronic resources

For specific sources, use of a search engine such as Alta Vista or Webcrawler will provide links to relevant sites and, since URL addresses change frequently, may be more up to date than the links provided below. Some useful examples are listed below, and provide a variety of links to related material:

The EMBL Nucleotide Sequence Database
http://mercury.ebi.ac.uk/ebi_docs/embl_db/ebi/topembl.html

National Center for Biotechnology Information, Genbank overview
http://www.ncbi.nlm.nih.gov/Web/Genbank/

PIR International Protein Sequence Database
http://www.bis.med.jhmi.edu/Dan/proteins/pir.html

SEQNET, SEQuence NETwork computer for molecular biologists
http://www.seqnet.dl.ac.uk/home.html

Swiss-Prot, annotated protein sequence database
http://expasy.hcuge.ch/sprot/sprot-top.html

Chapter 7

New and emerging viruses

The idea of 'new viruses' has attracted a great deal of public attention in recent years, but few such viruses are truly novel, and most are more accurately qualified as 'emerging'. An emerging disease is defined by the Centers for Disease Control (CDC) as:

> '*A disease of infectious origin with an incidence that has increased within the last two decades, or threatens to increase in the near future*'.

Unsurprisingly, while many diseases fit this definition, attention has tended to concentrate on viruses which cause incurable, fatal diseases such as hemorrhagic fevers or AIDS. In general, the term 'emerging disease' is most commonly applied to diseases which have historically been rarely seen (or identified). It is easier to see a change in a disease that has been restricted geographically or numerically. Indeed, even a major increase in an already common infection may be thought of as an epidemic or a pandemic rather than an 'emerging disease'.

Despite the entirely justified attention given to diseases such as the hemorrhagic fevers, many less spectacular diseases also fit the CDC definition. In fact, it is not even necessary for an emerging disease to be caused by a rare or novel virus, as the example of chickenpox in adults shows. This was traditionally an uncommon disease in temperate climates, but is becoming far more common as infection rates in childhood decrease. Unfortunately, adult chickenpox is far more severe, and the increase in the disease is a cause for concern. Despite the emerging nature of the disease, the virus itself is neither novel nor rare.

It is clear that the emergence of infections is not new. It has been occurring throughout history, and covers the whole range of infections, from the trivial (and often unnoticed) up to the high-profile killers.

7.1 Where do new or emerging viruses come from?

There are a number of routes by which viruses may 'emerge', and these are summarized in *Table 7.1*. It should be noted that these do not work in

Table 7.1: Routes for the introduction of novel viruses

Route	Example[a]	Contributory factors[a]
Zoonosis	Hantavirus, Dengue fever	Exposure to host species, habitat destruction, movement of animals
Mutation	Drug-resistant viruses influenza antigenic drift	RNA genome, poor use of antiviral drugs
Recombination	Influenza antigenic shift	Segmented genomes, exposure to animal sources of related virus
Geographical contact	Smallpox, measles	Population movements, trade routes, military expeditions
Deliberate release	Myxomatosis, smallpox	Pest control, bioweapons
Accidental release	Rabbit hemorrhagic disease, smallpox	Insufficient precautions
Genetic manipulation	Vaccinia recombinants, baculovirus recombinants	Economic and public health factors
Reappearance	(Smallpox)	Exhumations, excavations
Identification	Hepatitis C, hantavirus	Improved diagnostic techniques
Medical procedures	SV40	Animal-derived medical products, xenotransplantation
Unknown, probable zoonosis	Marburg, Ebola, HIV, BSE/nvCJD	Unknown (contact with host species?)

[a]Discussed in Section 7.1.

isolation. For example, a zoonotic infection may be significant only when humans are exposed to the normal host species by population movements, or may only become pathogenic when a mutant form arises.

7.1.1 Zoonosis

A zoonosis is the transfer to humans of an infectious agent from an animal source, and represents a major route for the introduction of novel pathogens. Examples of viral zoonoses are shown in *Table 7.2*.

A true zoonosis is poorly transmissible between humans, since otherwise human-to-human transmission comes to dominate spread of the virus. A possible example of this is HIV, for which a zoonotic origin (transmission from an African monkey to humans, based on similarities to SIV, the simian immunodeficiency virus) has been proposed. Clearly, even if HIV infection was originally a zoonosis, it is now a transmissible disease within the human population. Many viruses are well adapted to co-existence with their normal host and cause a relatively mild disease. As a result, a virus may circulate unnoticed in the animal population. However, if the virus is also able to infect humans, in some cases the resultant disease may be much more severe. Such a transfer requires contact between humans and animals, which may be increased by environmental changes.

Table 7.2: Examples of viral zoonoses

Disease	Frequency	Distribution	Reservoir host	Source of human infection	Virus family
Yellow fever	Endemic	Tropics	Monkeys	Vector (mosquito)	*Flaviviridae*
Colorado tick fever	Endemic	North America	Rodents	Vector (tick)	*Togaviridae*
Dengue fever	Endemic	Tropics	Monkeys	Vector (mosquito)	*Flaviviridae*
Lassa fever	Endemic	West Africa	Rats	Rats	*Arenaviridae*
Rabies	Endemic	World-wide	Mammals	Mammals (saliva)	*Rhabdoviridae*
HPS/HFRS[a]	Common	World-wide	Rodents	Rodents (urine, feces)	*Bunyaviridae*
Herpes B	Very rare	World-wide (laboratories)	Monkeys	Laboratory monkeys (bite)	*Herpesviridae*
Equine morbillivirus	Extremely rare	Australia	Unknown	Horse (body fluids)	*Paramyxoviridae*

[a]See Section 7.1.

A good example of this is Sin Nombre ('Without Name') virus. This was originally known as Four Corners virus, or Muerto Canyon virus, after the area in which it was first seen. The name change reflects an increasing desire to avoid 'geographical' naming of viruses, since it can be confusing when a virus named for one location appears somewhere different, possibly thousands of miles away.

In May 1993, a cluster of cases of a severe respiratory infection was noted in the 'Four Corners' area of the USA, where Arizona, Colorado, New Mexico and Utah meet. While the initial numbers were relatively small (12 deaths from 24 cases in the first half of 1993), mortality was extremely high, and would later rise to 80% in the initial patient group. The occurrence of multiple deaths in a relatively small area and in otherwise healthy individuals resulted in recognition of the new disease. Given the severity of the disease, finding the cause was a matter of urgency. By early June 1993, it was determined that the causative agent was a hantavirus. The first of this group (genus *Hantavirus*, family *Bunyaviridae*) to be identified was the Hantaan virus, noted during the Korean War as the cause of a hemorrhagic fever with renal syndrome (HFRS) in United Nations troops. In the case of the Korean virus, the condition had been known locally for some time (and fits with earlier historical reports), but only became known to Western medicine because of the presence of UN troops in the area. Further studies showed Hantaan-like HFRS disease in many countries throughout Asia and Europe.

With Sin Nombre, the disease was clearly different, but knowledge from the Korean outbreak was extremely valuable. It had been established that Hantaan was transmitted to humans by inhaling materials contaminated with the urine or feces of the animal host, the Asian striped field mouse (*Apodemus agrarius*). In the case of Sin Nombre, the animal host was quickly identified as the deer mouse (*Peromyscus maniculatus*). High snowfalls in the Four Corners area in the winter of 1992–1993 relieved a long-standing drought and, in the fertile spring that followed, the mouse population increased dramatically. As a result, people were increasingly exposed to their droppings and urine, and the disease, now termed hantavirus pulmonary syndrome (HPS), emerged. Interestingly, local native folklore speaks of a 'drowning sickness' in fertile years, suggesting that HPS may also have been observed historically. In fact, assessment of other cases of acute respiratory distress syndrome (ARDS) showed that HPS occurs all across the Americas, while HFRS seems to result from hantavirus infection in Europe and Asia. Hantaviruses are carried by a wide range of rodents, including mice, rats, voles, shrews and lemmings. In the rodent host, hantaviruses seem not to cause significant sickness, and indeed many hantaviruses are not associated with human disease. However, the severity of HFRS and HPS highlighted the role of hantaviruses in human disease, and it is suggested that many unexplained sicknesses that fit into these general categories could in fact represent unidentified hantavirus infections.

Hantaviruses illustrate many of the elements contributing to a zoonotic virus infection. Hantaan and Sin Nombre cause only inapparent disease in the animal host, in which they maintain a high level of infection, and which acts as a reservoir for human disease. They cause severe disease in humans, but are poorly transmissible. They were introduced into the human population by contact with the natural host, which increased due to environmental (Sin Nombre) or behavioral (Hantaan) changes. Despite this, Sin Nombre has been responsible for significant human disease in a highly developed country with excellent health surveillance, without being identified until the Four Corners outbreak brought it to the attention of the epidemiological community. Clearly there are likely to be many more such infections awaiting identification.

Although, as noted above, a zoonotic infection may spread directly from a mammalian species to man, transfer of many zoonotic infections involves carriage by another 'vector' species, and this is often an arthropod such as a mosquito or a tick. The virus is acquired when the vector feeds on the blood of the host. While it is possible for the virus to be transferred passively if a human is bitten soon afterwards, more commonly, the virus may replicate in the vector species. In some cases, the virus may even be able to be passed to subsequent generations of the vector in the eggs. A functional classification of such diseases as arthropod-borne viruses (arboviruses) has been widely used. There are well over a hundred arbovirus infections of humans, and it has been suggested that at least 20 of these can be classified as emerging diseases.

Dengue hemorrhagic fever (*Flaviviridae*) is a major example of an arthropod-borne viral zoonosis. It is endemic to tropical and sub-tropical areas world-wide, and is transferred between monkey hosts by *Aedes* mosquitoes. In humans, exposure to one serotype (serological variant) of the Dengue virus results in Dengue ('breakbone') fever, characterized by high fever and muscle and joint aches. However, such infection does not give immunity to other serotypes. Rather, the disease caused by subsequent infections with another serotype is made far more severe, involving life-threatening hemorrhagic disease. This is because the (non-neutralizing) circulating antibodies resulting from the earlier infection assist the virus in getting inside monocytes (see Chapter 2). Once inside, however, rather than being digested, the virus replicates in these cells and causes Dengue hemorrhagic fever, in which the course of the infection is complicated by capillary leakage leading to hemoconcentration and consequent shock, with significantly increased mortality. Clearly, this disease relies on the presence of multiple serotypes of the virus in a population, and until approximately 50 years ago the different serotypes were restricted to different geographical areas. It is thought that the population movements (civilian and military) and loss of effective hygiene during World War 2 were responsible for spreading these previously isolated serotypes among the human population.

Dengue is carried by mosquitoes, in particular *Aedes aegyptii*, although the risk of spread of the virus was significantly increased in the United States when another mosquito host species, the aggressive Asian 'tiger mosquito' *Aedes albopictus*, was introduced to the Americas in 1985 in water contained in a shipment of used car tires. Dengue hemorrhagic fever illustrates the roles of many factors in an emerging viral zoonosis, including population and animal movements, as well as the properties of the virus itself.

One question that must be asked is: why is a zoonotic infection often more severe than in its original host? Although many zoonoses do not appear to cause significant disease (and may as a result not be noticed), it is clear that many cause severe infections in the new host while being relatively harmless in the normal host. There are alternative explanations for this.

The classical theory states that it is in the interest of a virus not to kill or immobilize the host too quickly, since this reduces the chances for spread of the virus. Over time, both virus and host adapt so that the virus produces less severe disease in its normal host. The moderation of virulence of myxomatosis in the Australian rabbit population (see Section 7.1.5) illustrates such adaptation. However, when a novel host is infected, there has been no chance for adaptation to a less virulent form, and severe disease may result, at least initially. More recent theories suggest that this is an oversimplification, and that such benign equilibrium is subject to disruption by more virulent agents and by methods of transmission that are relatively unaffected by the sickness of the host (e.g. arthropod vector transmission). However, it should be noted that myxomatosis is primarily transmitted by arthropods, but still provides a classical example of benign adaptation.

A more extreme form of the benign adaptation hypothesis has also been suggested (the 'virus X' theory) in which viruses may act as a symbiote (providing mutual benefit), providing defense against 'invaders' rather than being parasites (damaging the host for their own benefit). Thus, while not harming the natural host, they could infect a new species with which the host comes into contact, assisting the normal host to win out in competition with the sickened opponent species. Such effects are even observed within one species (see Section 7.1.4). This theory fits well with the many monkey and ape viruses that can infect and cause severe disease in the closely related human species; it is not long in evolutionary terms since we were just another African plains ape. However, there are some problems with the theory. Firstly, many viruses are still highly virulent in, and specific to, the normal host, and is difficult to see the evolutionary advantage of a virus such as hepatitis B. Secondly, many viruses do not cause disease, or cause only mild disease in other species (e.g. many of the hantaviruses). Thirdly, it is difficult to see the evolutionary advantage conferred upon some species by the zoonoses for which they are responsible. For example, Sin Nombre is unlikely to play a significant role in protecting deer mice from humans. It is clear that not all viruses are 'virus X' agents, but effects of this type may play a role.

As with all biological systems, it is likely that a combination of factors are responsible for the observed effects of zoonotic infections, and that no single theory will explain all of the observed events. What is clear is that zoonoses represent a major source of severe infections in humans.

7.1.2 Mutation

While all viruses must at base have arisen by mutation, the effects of this process in the limited time that we have been able to study viruses are necessarily limited. Historically, there are reports of apparently novel diseases, but the reliability of such observations is doubtful given the rudimentary state of medical knowledge at the time. In addition, the records used to support such findings are fragmentary and cover very limited areas, so it is entirely possible that a similar disease was active, but unrecorded, in other areas. More recently, there is evidence of new diseases arising, most notably in the case of AIDS, where despite the present pandemic, serological evidence indicates that the disease was not present in the general human population before the second half of the twentieth century (see Section 7.1.4), and with bovine spongiform encephalopathy (BSE, see Sections 1.8.6 and 7.1.10), although since the BSE agent is thought to be a protein-only infectious agent, mutation in the classical sense could not be involved. The appearance of rabbit hemorrhagic disease calicivirus (see Section 7.1.6) also represents an apparent 'new disease' followed by rapid and devastating spread.

One example of ongoing virus mutation which can be observed is antigenic drift in influenza virus (see Section 1.4.5). However, the mutant forms produced are not easily resolved from the parental strains, and are not thought of as new or emerging viruses. A more relevant example is the appearance of drug-resistant variants of almost all viruses that are treated with antiviral drugs (see Chapter 4). While drug resistance arises by mutation, this does not have to occur after the drug is used. It has become clear that even in any one host, mutation during normal replication results in the presence of a population of viruses with differing genetic and physical properties, rather than a single type. Some of these will be drug-resistant mutants, which may be at a disadvantage normally, but will be preferred when the antiviral drug to which they are resistant is used (see Section 4.3). The problem of drug resistance is particularly marked in bacteria, where the genes responsible may be transferred on extrachromosomal genetic elements (plasmids), even between different species. Drug-resistant bacteria such as methicillin-resistant *Staphylococcus aureus* (MRSA) or multidrug-resistant *Mycobacterium tuberculosis* are often referred to as emerging diseases in their own right, and drug-resistant viruses may yet achieve this status. It is to be hoped that lack of equivalent resistance transfer mechanisms in viruses and the judicious use of antiviral drugs will slow the appearance of viral equivalents. However, there are already reports of forms of HIV resistant to multiple drugs of the same general type, and the rapid mutation of viruses makes this likely to become more of a problem in future.

Mutation of viruses is the key to escaping immune surveillance, as well as resisting the effects of antiviral drugs. As stated above, few of these viruses are sufficiently different to be classified as emerging diseases. Despite this, mutation does contribute to emerging diseases. For example, hepatitis C is showing a dramatic increase in frequency at least in part due to the extreme antigenic variation of the virus, the result of a high mutation rate.

7.1.3 Recombination

While it is possible for any virus to undergo recombination, where genetic material is exchanged with related viral or cellular sequences, this is most evident in influenza virus, where the RNA genome is present in multiple different segments which can be exchanged without any need to cut and splice nucleic acids. It is known that influenza, as well as undergoing mutations of the surface glycoproteins hemagglutinin (HA) and neuraminidase (NA) (antigenic drift), can also undergo sudden changes of these proteins. This is thought to result from the exchange of genome segments with animal influenza viruses during mixed infections, often as a result of the close proximity of humans, farm animals and birds (all of which can act as hosts), and is known as antigenic shift (see Section 1.4.5). Immunity to the original human virus does not protect against the shifted virus, which can then cause the world-wide infection known as a pandemic. There have been five pandemics this century. In the most severe of these, the 1918–1919 'swine flu' pandemic, over 20 million people died (up to 100 million by some estimates). The virus that caused this disease resembled a pig influenza virus in the structure of the HA surface glycoprotein, although the other surface glycoprotein (NA) was apparently of human origin. Influenza viruses are described by the serotype of their surface glycoproteins, which are given numbers. Swine flu virus was termed $H_{SW}1N1$, and viruses of this type have not been seen since the 1918–1919 pandemic. An $H_{SW}1N1$ influenza was seen at Fort Dix, New Jersey in 1976 and resulted in a nationwide vaccination program, but the virus proved to be very different from the 1918–1919 virus, in that it was poorly transmissible and caused only a very limited local outbreak. A total of 13 cases of $H_{SW}1N1$ influenza (or co-infections) were reported, with one fatality. Although viruses produced by antigenic shift are still referred to as influenza, they are different enough for the immune system to regard them as new viruses. Recently, an influenza virus similar to avian strains, with a novel serotype referred to as H5N1, has caused human infection in Hong Kong, and there is great concern that this could represent the start of a pandemic.

While all viruses can undergo recombination, the presence of a segmented genome makes this more likely since no molecular changes are required. There are many other viruses with multiple genome segments, including the *Reoviridae* and the *Bunyaviridae*. The former include rotaviruses, which are a major cause of infant mortality in the developing world, while the latter include the hantaviruses discussed in Section 7.1.1.

7.1.4 Geographical contact

In the predominantly agricultural society with little travel between regions that existed for most of recorded human history, it is possible for viruses to exist in isolated areas of infection. However, as society develops, trade routes are established that allow for the transmission of infections between such areas. As more and more people travel, such transmission becomes more common along the trade routes. Military expeditions can also be responsible for spreading infection. It has been suggested that the ability of the Spanish *conquistadores* under Cortes to conquer the complex and populous Aztec nation was due more to European viral infections (including smallpox and measles) passed to the totally non-immune native population than to military advantages such as gunpowder. Certainly the death toll from such introduced diseases was very high, with some areas being effectively depopulated. Dengue hemorrhagic fever is another disease where population movements due to military action have played a significant role, as discussed in Section 7.1.1.

While a zoonotic origin for AIDS seems likely, it is also probable that the virus may have been present in restricted areas of central Africa for some time. Given the lack of epidemiological data on such areas, the virus would not have come to the attention of the medical community while it was restricted to this region. The opening of central African regions to commerce, notably by the building of the Kinshasa highway across the region, is likely to have played a major role in such spread. The factors involved in amplifying the disease once it was established in the wider population are discussed in Section 7.2. It should be noted that population movements into new rural areas provide an important route for the introduction of zoonotic viruses. Even the most efficient potentially zoonotic 'new virus' will not be able to infect a human if the host or carrier species does not come into contact with any. However, settlers in new areas may come into increased contact with potential animal hosts or carriers. In addition, the resultant disruption to the life of the animal hosts may also play a role. If a virus is circulating efficiently in a large animal population, there is little selective pressure for the virus to evolve the ability to infect another host species, particularly if this incurs a decrease in the ability to replicate in the normal host. However, if the numbers of the normal host are reduced (e.g. by deforestation), circulation of the virus will be disrupted and a virus which can infect another host may be at a selective advantage, even if it has lost some ability to replicate in the normal host.

It should be noted that it is not just human population movements that can introduce diseases. International travel is usually thought of as a human phenomenon, but movement of agricultural and experimental animals is also a significant factor. Ebola Reston strain was introduced to America in a shipment of monkeys from the Philippines. The monkeys died of the infection, but fortunately this particular strain proved not to cause human disease. Spread of animal viruses is in some ways easier to control. Transfer of animals is generally more regulated than human travel, and it is possible to introduce programs of testing, quarantine and (on occasion) mass slaughter that would be unacceptable with humans. In the early history of the AIDS pandemic, many countries (including some with significant levels of infection already present within their borders) introduced requirements similar to the milder forms of animal disease regulation, in that they required that travelers be certified as HIV-negative before entering the country. Even these mild controls proved very unpopular, and they were generally short lived.

With international travel at its present level, few areas remain where isolated diseases can circulate. Despite this, some concerns about this remain. Recently, attention has focused on American troops returning from military training exercises in areas of Australia where the mosquito-borne Ross River virus (*Togaviridae*) is endemic. There is serious concern that this disease could be introduced into America, and extensive control measures (aimed particularly at vector control) were taken before, during and after the exercise to prevent this. Despite these, at least one case was identified (and contained) in the United States following the exercise.

An extreme form of the 'geographical contact' route has been suggested based on the identification of organic materials in cometary matter. The suggestion that 'viruses from outer space' invade this planet is extremely hypothetical to say the least. Influenza was originally so named in Renaissance Italy since the disease was thought to be due to the influence of the stars, but this does not constitute proof.

7.1.5 Deliberate release

The best known example of the deliberate release of a pathogenic virus is the use of the myxoma poxvirus against the Australian rabbit population. In American rabbits (the normal host), this virus causes mild and self-limiting skin disease, but in European rabbits it causes myxomatosis, a severe systemic disease lethal to more than 90% of infected rabbits. In a strikingly unsuccessful attempt to provide a useful economic resource, a limited population of European rabbits was introduced to Australia in the nineteenth century. Escaped rabbits multiplied rapidly, and by the twentieth century they were a massively destructive and costly pest. Myxoma virus is spread readily between rabbits by mosquitoes and fleas, and was released into the Australian rabbit population in 1950 in an attempt to control their numbers. In the short term it was highly successful, but within a few years the majority of rabbits were resistant to the virus due to 'natural' selection of disease-resistant rabbits along with the evolution of less pathogenic viruses which allowed their hosts to survive longer (particularly over the winter, when mosquitoes and fleas are less common) and so to spread the virus more effectively. As a result, a general immunity to the virus was established, so that within a decade the rabbit population was recovering rapidly. As a result, other agents have been investigated for use in Australia, including the calicivirus responsible for rabbit hemorrhagic disease (see Section 7.1.6). It now appears that this agent has also been deliberately and illegally released within New Zealand, although this has since been authorized 'after the event' by the New Zealand government.

It is sadly true that biological weapons have been used against humans as well as in agricultural control. While some cases were the unplanned result of contact between disparate groups (see Section 7.1.4), deliberate releases have occurred. Smallpox in particular has been used in this way, by supplying contaminated blankets to Native American tribes in the early years of the colonization of the American West.

In June 1763, there was an outbreak of smallpox at Fort Pitt on the Ohio River, which was under siege by the Shawnee, Delaware and Mingo tribes. In July, following correspondence between Colonel Bouquet, the commander of the fort, and Lord Jeffrey Amherst, the British military commander in North America, it was decided to "try to inocculate [sic] the Indians by means of Blanketts [sic] that may fall in their hands" (*Figure 7.1*). The attempt to 'innoculate' succeeded, and the resultant epidemic spread across the continent, killing many thousands of Native Americans from at least six tribes, as well as substantial numbers of colonists.

Although there are now extensive international treaties in place to prevent the development or use of biological weapons by governments, active research toward this end was undertaken on virulent human pathogens in some countries at least into the 1970s. Although no biological weapon of this type has been used, the possibility of release by terrorists or other sources continues to cause concern.

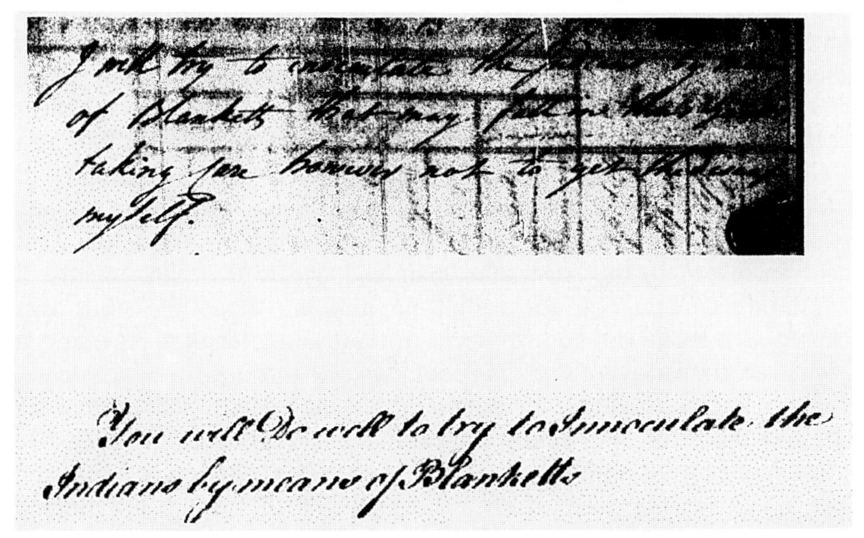

You will Do well to try to Innoculate the
Indians by means of Blanketts

Figure 7.1: Correspondence between Colonel Henry Bouquet and Lord Jeffrey Amherst. Colonel Bouquet writes. "I will try to inocculate the Indians by means of Blanketts that may fall in their hands, taking care however not to get the disease myself". Lord Amherst replies: "You will Do well to try to Innoculate the Indians by means of Blanketts". Documents reproduced from web site http://www.nativeweb.org/pages/legal/amherst/lord_jeff.html with permission from Peter d'Errico, Legal Studies Faculty, University of Massachusetts, Amherst, MA, USA.

One other area where the use of viral agents as controls is being actively investigated is the use of bacteriophages as therapeutic agents in drug-resistant bacterial disease. Some of the earliest work with bacteriophages, published in 1917, suggested that viral killing of dysentery bacteria was responsible for controlling human disease. In the following two decades, there was a great deal of interest in the possibility of 'bacteriophage therapy' for this and other bacterial diseases. This was generally unsuccessful, and interest in the field waned rapidly with the development of effective antibiotics in the 1940s. However, since the appearance of antibiotic resistance in bacteria, interest in bacteriophage therapy has revived. Although there are problems with the induced immune response to the bacteriophages themselves and with clearance of the bacteriophage from the blood, we now have a far greater understanding of the biology of bacteriophages (and the possibility of genetically manipulating the bacteriophage genome), and it seems likely that practical applications will be found for this approach.

7.1.6 Accidental release

Once again, the best example of this is provided by Australian efforts to control the rampant European rabbit population in their country. Rabbit hemorrhagic disease is caused by a member of the *Caliciviridae*. It was first noted in China in 1984, and spread across Europe and Asia, causing up to 95% mortality in the rabbit population. In 1995, evaluation of the virus for biological control began on Wardang Island, several miles off the coast of South Australia. There was concern regarding the susceptibility of native Australian species to the virus, and it was being tested in an isolated setting before any release. By early 1996, via an unknown route, the virus was transmitted to the Australian mainland, where it began to cause a devastating epidemic in the rabbit population. While the virus was licensed as a biological control agent in Australia later that year, and is now in use, the escape of the virus does illustrate the hazards of accidental release. It is estimated that 20 million rabbits died within 3 months of the release of this virus. It does now appear that concerns regarding disease in native species were unfounded, but these species are suffering indirectly, since predators deprived of their normal diet of rabbits are finding alternative food supplies. This is a major concern following the illegal release of this virus in New Zealand (see Section 7.1.5).

Once again, smallpox provides an example of accidental release in the human population. Vaccination against this devastating disease had been available since Benjamin Jesty and Edward Jenner pioneered the use of cowpox as a vaccine in the late eighteenth century. Following a concerted eradication campaign begun in 1958, the last 'wild' case of the variola major form of smallpox occurred in 1975, and the last case of the milder variola minor form in Somalia in 1977. It was decreed that a period of 10 years was required before smallpox could be declared eradicated. However, in 1979, two cases of smallpox occurred (with one death) following a laboratory accident at the University of Birmingham, UK. The outbreak was contained to just these two cases, and smallpox is now officially eradicated. The last remaining stocks of the virus (held in Novosibirsk, Russia and at the Centers for Disease Control in Atlanta) are now scheduled to be eliminated on the 30th of June, 1999, at least in part due to the events in Birmingham.

7.1.7 Genetic manipulation

With the routine use of recombinant DNA technology, it is now theoretically possible to produce a virus which expresses almost any desired protein. Baculoviruses are viruses of insects that are widely used as cloning vectors, but baculoviruses are also highly species-specific and virulent pathogens of a wide range of insect species. Late in infection, many baculoviruses produce massive amounts of a specialized protein (polyhedrin in some, granulin in others), which forms a protective capsule around the virus. This structure, consisting of virus(es) and capsule is referred to as an occlusion body and makes virus infectivity very resistant to environmental degradation. However, if the occluded virus is eaten by a caterpillar, the occlusion body breaks down and releases the virus into its gut, initiating infection. This combination of environmental stability and lethality to destructive pests meant that baculoviruses were in themselves promising biological insecticides. Work with occluding baculoviruses has led to the marketing of a multiple baculovirus preparations for the control of a wide range of crop-destroying insects. However, while normal (non-recombinant) baculoviruses can be highly effective insecticides, they tend to be slower acting than chemical insecticides and, as a result, there has been a great deal of interest in the introduction of genes intended to make baculoviruses more rapidly lethal to their insect hosts. Such genes include proteins to interfere with insect hormone function as well as insect-specific toxins. In some cases, the effect of these toxins is similar to that of chemical insecticides, with the major advantage that they are produced within (and only within) the insect itself. Extensive testing and trials have been undertaken, and it is very unlikely that any such virus will threaten humans, but it is clear that for the insect target they will represent an significant new disease.

Other uses of genetically manipulated viruses are as vaccines or immunotherapeutic agents (see Chapter 2), as agents for gene therapy (see Chapter 5) and as targeted cytopathic agents in cancer therapy. A wide range of such viruses are in testing. As vaccines, they may be deleted versions of the parental strain, with essential genes missing (essentially acting as defective viruses rather than new viruses), but for some vaccination approaches and all gene therapy approaches, additional proteins are expressed. The virus vectors used are modified so that they are very unlikely to be transmitted, and extensive testing is carried out on any recombinant agent. In consequence, although they are in fact 'new viruses', they are very unlikely to cause disease. Anticancer applications are often similar to gene therapy, relying on localized expression of a gene with cytopathic effects (virus-directed enzyme–prodrug therapy and other approaches).

One example of a recombinant virus already in use is a vaccinia poxvirus expressing the G glycoprotein of rabies virus within the vaccinia thymidine kinase gene (see Sections 3.3 and 3.6). Using this location for inserted DNA reduces the virulence of the vaccinia vector. The vaccine is used for preventing rabies in wild animal populations. It is given inside an edible capsule, and is thought to enter the bloodstream via cuts in the mouth. The recombinant vaccine has been in use in the field since 1987 with no significant health concerns, and appears to be effective in controlling rabies in wild fox and racoon populations.

7.1.8 Reappearance

Clearly, for a virus to reappear, it must have disappeared in the first place, and few viruses meet this requirement. One such virus is smallpox. Monkeypox virus is a close relative of smallpox, and recent rises in transmission are thought to be due to decreased levels of smallpox vaccination. However, monkeypox remains to date a limited, non-epidemic disease. Initial concerns that monkeypox was actually a form of smallpox were resolved when full genetic sequences of the two viruses were obtained, and it became clear that there were substantial differences at the genomic level. Another potential route does remain, since poxviruses are very stable, and there is still some concern that building works in long-established cities (which often break open previously unknown grave sites) or deliberate archeological investigations, could result in release of live virus. While the virus appears to lose at least some activity within months in warm conditions, it is stabilized within scab tissue from victims of the disease and at colder temperatures. Morphologically intact virus has been observed from mummified tissue 500 years old, although the virus was not viable. It should be noted that viruses of humans do not form environmentally stable spores as do some bacteria, although the occluding baculoviruses of insects have an analogous mechanism (see Section 7.1.6). However, there is sufficient concern about survival of smallpox virus in scab tissue within closed burial caskets in cool climates that vaccination of workers coming into contact with such tissues has been suggested.

One other candidate for reappearance is the 'swine flu' ($H_{SW}1N1$) variant of influenza responsible for the 1918–1919 pandemic. Despite a scare at Fort Dix, New Jersey in 1976, when a related but less virulent virus was observed, the pandemic form of swine flu has not been seen for almost 80 years. Exhuming the remains of victims of this epidemic from burial sites in permafrost is being attempted in order to carry out molecular studies. While social disruption associated with World War 1 is thought to have been responsible for increasing the casualty rate resulting in influenza in 1918–1919, and also while it is very unlikely that the virus (which is relatively fragile) could survive in an infectious form for so long, even in permafrost, it is impossible to discount the risk entirely. The precautions taken will of course reflect this concern.

7.1.9 Identification

While we now know of many hundreds of viruses infecting humans, it is clear that many more remain to be discovered. Many of these will cause only mild disease, since those which cause obvious symptoms tend to be the first to be noticed. However, there are also a range of 'orphan' diseases for which a cause has not yet been found. Molecular techniques have greatly extended our ability to determine the nature of the agent in such cases.

In the case of hepatitis C virus, a 'virus fraction' was prepared from plasma containing the infectious agent for 'non-A, non-B' (NANB) hepatitis. A clone prepared from this was expressed, and the resultant protein found to react with patient sera. This allowed diagnostic tests to be developed. In subsequent work, a virus was observed, but this did not occur for some time. The virus has been classified as a member of the *Flaviviridae* (see Section 1.2) and is clearly a major agent of human disease. In fact, it appears that the incidence of hepatitis C infection is increasing, and it is now being referred to as an emerging disease.

With human herpesvirus 8, which appears to be involved with Kaposi's sarcoma (KS) in AIDS patients as well as with other malignant conditions, the viral genome was detected by 'representational difference analysis'. DNA from a KS lesion was combined with DNA from a sample taken from elsewhere on the same patient. The DNA that did not react, and was therefore present only in the KS lesion, was examined, and sequences were found which were clearly those of a member of the *Gammaherpesvirinae*, related to the oncogenic virus of squirrel monkeys, Herpesvirus saimirii, and more distantly related to EBV (human herpesvirus 4), which causes a variety of human cancers (see Section 1.7). KS was traditionally a rare and relatively benign cancer. However, both the incidence and the severity of the disease have increased due to immunosuppression caused by HIV, illustrating another route for disease emergence.

Both of the above viruses have proved extremely difficult to grow in culture, and would not have been found without the use of molecular techniques. Clearly, identification of a pre-existing virus does not make it either new or emerging, but it does allow studies to identify the role and prevalence of the virus in question. With both of the viruses discussed above, such studies showed that the virus was both a major health problem and also increasing in frequency, classifying the newly identified virus as an emerging disease.

7.1.10 Medical procedures

Any procedure where animal-derived biological materials are introduced into the body carries with it the risk of introducing a zoonotic infection. In the case of some such materials, such as vaccines, extensive purification and inactivation procedures may be followed. In the case of others, most notably organ transplants, it is not possible to perform such procedures. While many viruses of humans can be transmitted in this way, they are not the primary concern of this chapter. Two examples suffice to illustrate the potential for new viruses to enter humans by this route.

The SV40 papovavirus is a naturally occurring macaque virus which was present in early polio vaccines due to their growth in rhesus monkey kidney cells (see Section 3.1). Fortunately, this was an example of apparently harmless zoonosis, since despite a great deal of anxiety, no human pathology has been confirmed as resulting from SV40 infection from polio vaccine in over 40 years. Many safeguards are now in place to prevent such incidents.

The possibility of xenotransplantation is now causing a great deal of discussion. In such procedures, organs from animals are used to replace failed human organs. While initial attention focused on the use of primate organs, notably those from the baboon, concerns about the apparent ability of baboon viruses to infect human cells caused interest to switch to transgenic pigs expressing human cellular marker proteins in order to minimize rejection. However, recent work has shown that at least one pig retrovirus, which has been named Circe, can infect human cells. Since retroviruses are routinely present in mammalian cells (and may be essential, see Section 1.6), it is not possible to prevent the introduction of such agents. While donor animals would be bred under gnotobiotic conditions (as 'germ-free' as possible), other viruses could also be present (including endogenous retroviruses transmitted within the animal DNA), and the ability of the pig to act as a 'pressure cooker' to recombine strains of influenza has been noted. At present, the options for xenotransplantation are being reviewed. Potential recipients often see it as a personal issue, with the possibility of future, unknown infections outweighed by the near certainty of death without a donor organ which may not be available from a human source. However, epidemiologists are very concerned about the possibility of introducing zoonotic infections directly into patients who are immunosuppressed to prevent rejection of the implant. These infections could then be 'amplified' (see Section 7.2) and introduced into the human population as a whole. It is likely that xenotransplantation will proceed, with very strict safeguards, but the risks of such procedures require thorough evaluation.

It should be noted that immunosuppression, whether as a medical procedure or due to other effects, could in itself allow novel mutants or poorly replicating zoonoses to establish and transmit to other humans. Increasing levels of immunosuppressive procedures, infections and other agents are, therefore, also a cause for concern.

7.1.11 Mysterious viruses

While the filoviruses Ebola and Marburg are often spoken of as zoonotic infections, and it is clear that there must be a reservoir host since there is no evidence of circulating virus in any human population (and given the nature of the infection this seems unlikely), the reservoir has not yet been identified. It is known that one strain of Ebola (Reston, which is not pathogenic in humans) is present in Filipino macaques, and monkeys may provide a route for transmission to humans for other strains. However, the virus appears to cause severe disease in monkeys, suggesting that they are also the victims of a zoonosis rather than a reservoir host. Many species have been studied (including some plants), and there is limited evidence that bats can carry the infection. However, studies on bats at the site of some outbreaks (notably N'Zara in the Sudan and Kitum Cave in Kenya) have proven inconclusive.

BSE is a member of the group of diseases known as transmissible spongiform encephalopathies (TSEs). Its appearance in cows in the UK in the mid-1980s was originally thought to be due to changes in rendering sheep carcasses used to provide protein for bovine feeds. This reflected the knowledge that sheep suffer from the TSE scrapie (see Section 1.8.6). However, although scrapie may be experimentally transmitted to rodents, it does not appear to cross into novel species very efficiently, and no human risk from scrapie has ever been identified. More recent data suggests that BSE may not be a zoonotic form of scrapie, but rather may have arisen spontaneously in the UK cow population, as the TSE Creutzfeld–Jakob disease (CJD) may do in humans (see Section 1.8.6). It would then have been amplified by the use of bovine materials in animal-derived feed ('feeding cows to cows'). What fuels most of the concern is that, unlike scrapie, there is increasing evidence of transmission of the BSE agent to humans, where it appears to take the form of 'new variant' CJD (nvCJD), which has a different clinical pattern and affects different age groups compared with classical CJD. Changes in meat preparation in the UK in the years following the identification of the disease significantly reduced the risk of exposure, specifically by banning the use of neurological tissue in food. The use of such material in prepared 'meat' products is controversial, but was widespread at the time, and is not yet banned in many countries. If the agent was transmitted directly from neurological tissue, it seems likely that the high level of exposure among slaughterhouse workers would lead to a high incidence in this population, but this has not been seen. The cases that have appeared may represent only the very start of an increasing curve, or alternatively may be the peak of a limited zoonosis. Infection, if it occurs, appears to be by consumption of beef and beef products, probably prior to the banning of neurological tissue in food products. There is also evidence that specific forms of the human PrP gene in sufferers (see Section 1.8.6) may be involved in progression to disease. Links to 'downer cow syndrome', widespread in the United States, have also been suggested. Despite intensive study, the origin and nature of BSE remain controversial and unresolved.

7.2 What factors contribute to disease emergence?

Once a 'new virus' has entered the human population, this does not necessarily make it an emerging disease, for a number of reasons:

- Many viruses do not cause significant disease in humans. Such viruses cannot truly be classified as emerging diseases, even if they become widespread. However, with improved diagnostic and epidemiological techniques, even previously unidentified agents which usually cause only mild disease may become known. Some of these agents may be very widespread, such as human herpesvirus 6 (HHV-6) which causes a mild and transient childhood exanthem (roseola infantum). However, there is always the possibility that such agents may become implicated in more severe disease, as appears to be happening with HHV-6.

- Poorly transmissible agents infect only limited numbers of people, and may remain unnoticed. Despite this, severe infections of this type, such as Ebola (transmission of which requires close contact), are often considered emerging diseases due to the level of concern about such devastating infections. Even though Ebola outbreaks to date have been geographically restricted and of limited duration, the nature of the disease and the possibility of spread within an urban environment cause genuine concern.
- Even if a virus establishes itself within the human population, it may remain restricted to a few hosts or one area. Examples of this include Dengue fever prior to World War 2, and may also include HIV. It is thought that HIV may have been present in rural areas of Africa for some time before transmission to urban areas and into 'amplifier' populations where more intense and high-risk sexual activities combined with injecting drug abuse converted it into a pandemic.

The role of specific factors in disease emergence is summarized in *Table 7.3*. It should be noticed that these factors rarely operate in isolation, for example war increases poverty, reduces hygiene, and can

Table 7.3: Factors favoring the emergence of infectious diseases introduced into the human population[a]

Factor	Effects
Poverty	Poor diet
	Poor hygiene
	Immunosuppression from other infections
Urbanization	Close proximity of hosts
	Poor hygiene
	Poor housing
	Relaxation of social codes
War	Disruption of medical services and disease control
	Movement of refugees and troops
	Overcrowding and poor hygiene in refugee camps
	Poverty
Irrigation	Creation of breeding pools for vector mosquitoes
Population movements	Spread of previously localized diseases and of infected carriers (or vectors)
Multiple partner sexual activity	Transmission of blood-borne and sexually transmitted agents (risk greatly increased by extreme practices)
Injecting drug abuse	Transmission of blood-borne agents
	Immunosuppression (direct and from other infections)
Medical procedures	Transmission of blood-borne agents by re-use of needles
	Transmission by blood transfusion, blood products or organ transplant
Immunosuppressive conditions[b]	Favor emergence of previously non-pathogenic agents as causes of significant diseases

[a]Combinations of these factors apply in many cases.
[b]May result from deliberate medical procedures (cancer therapies, immunosuppression for organ transplantation), environmental pollutants (e.g. organochlorines in seawater) or infectious agents (e.g. the human immunodeficiency virus, HIV).

alter the pattern of sexual activity in an area. Specific examples of viral disease amplification in the human population include the following.

- The endemic status of hepatitis A and E in urban areas with poor sanitation where fecal contamination of drinking water is unavoidable.
- The rise in almost all blood-borne diseases (notably hepatitis B and C) resulting from use of shared needles in injecting drug abuse, combined with the poverty and poor hygiene of many drug abusers and possible direct immunosuppressive effects of some drugs of abuse.
- The explosive increase in HIV infection resulting from extreme promiscuity and specific practices (notably, unprotected receptive anal sexual intercourse) in homosexual populations in the early stages of the HIV pandemic. In counterpoint to this, population movements within Africa, in particular the displacement of male workers from their families for long periods, has led to a far greater problem of heterosexual AIDS in such countries, with prostitutes as major vectors.
- The increase in the number of infections in hemorrhagic fever outbreaks resulting from centralized care of victims in hospitals with very limited facilities. The re-use of needles in such hospitals is thought to be a major route of spread. Additionally, since carers in such hospitals do not have access to effective isolation procedures, they may become infected and can then further spread the virus. Re-use of needles has also been implicated in the spread of HIV both in Africa and in Eastern Europe.

It can be seen that while diseases must first be introduced into the human population, a combination of other factors is required for them to become established and transmitted within that population. It is probable that many agents with the potential to cause significant human disease have failed at this stage, and the eradication of the risks outlined in *Table 7.3* plays an important role in disease control.

7.3 Monitoring emerging infectious diseases

It is becoming increasingly clear to all concerned that monitoring the appearance and frequency of infectious diseases is an essential part of disease control, and that surveillance in developing countries is an essential part of any such program. The US Senate Foreign Operations Subcommittee recently noted that "There is an urgent need . . . . to significantly augment international surveillance and control mechanisms, and to strengthen the ability of developing countries, where deadly viruses often first gain a foothold, to protect and care for their people".

The process of identifying new or emerging diseases involves work at all levels, from the 'disease cowboy' trying to find an outbreak, whose biggest problem may be where to find spare parts for his jeep, to the molecular biologist compiling sequence data who may know almost nothing about the source of the disease. When a mistake is made, it is easy to apportion blame with the benefit of hindsight, but it is a lot more difficult to know where to use the necessarily limited resources available before an outbreak has occurred. Two examples illustrate the opposing problems of too little and too much.

The outbreak of Sin Nombre virus discussed in Section 7.1.1 was detected by alert epidemiologists, and since the identification of Hantaan virus during the Korean war the United States Army had run a program investigating hantavirus disease. Unfortunately, budget cuts in 1991–1992 had greatly reduced this program, just before the 1993 outbreak. Of course, at the time, there was no sign that hantaviruses would cause any major problems in the USA, and cutting the budget seemed 'harmless'.

At the opposite extreme is the consequences of the 1976 Fort Dix swine flu outbreak (see Section 7.1.3). The virus isolated in an outbreak at this military training facility appeared by all available tests to be very closely related to the pandemic 1918–1919 'swine influenza' which killed more than 20 million people. A massive public health program was begun to head off the expected epidemic, which some predictions suggested could kill a million Americans. Almost 50 million Americans were vaccinated, and huge efforts were made to ensure that there was sufficient vaccine for every American. Had the epidemic developed as expected, this would have been hailed as a triumph of public health medicine. However, the virus proved to be poorly transmissible and caused only a localized outbreak. In consequence, the main results of the vaccination program were substantial damage to the public perception of vaccination, together with a series of claims for damages due to the vaccine, most notably involving Guillain–Barre syndrome, a paralytic neurological condition. Settling the claims would eventually cost nearly US$100 million, almost as much as the vaccination program itself. Yet, under different circumstances, canceling the hantavirus programs could have been harmless, and the swine flu vaccine could have saved millions of lives. While making such decisions is complex, and unlikely ever to be right all the time, the cost of the wrong decision can be very high indeed.

Further reading

Bonning, B.C. and Hammock, B.D. (1996) Development of recombinant baculoviruses for insect control. *Annu. Rev. Entomol.*, **41**, 191–210.

Cory, J.S. (1991) Release of genetically modified viruses. *Rev. Med. Virol.*, **1**, 79–88.

Ewald, P.W. (1997) Guarding against the most dangerous emerging pathogens: insights from evolutionary biology. *Emerging Infect. Dis.*, **2**, 245–256.

Garrett, L. (1994) *The Coming Plague*. Penguin Books, New York.

McNeill, W.H. (1976) *Plagues and Peoples*. Doubleday, New York.

Ryan, F. (1997) *Virus X: Tracking the New Killer Plagues Out of the Present and into the Future*. Little Brown & Company, New York.

Electronic resources

For specific sources, use of a search engine such as Alta Vista or Webcrawler will provide links to relevant sites and, since URL addresses change frequently, may be more up to date than the links provided below. Some useful examples are listed below:

Emerging Infectious Diseases Online
http://www.cdc.gov/ncidod/EID/eid.htm

Promed mail: Programme for Monitoring Emerging infectious Diseases Email discussion forum. Offers full contents or digests
Email to majordomo@usa.healthnet.org, subject 'subscribe promed' or 'subscribe promed-digest'

Outbreak Online
http://www.outbreak.org/cgi-unreg/dynaserve.exe/index.html

Appendix A. Glossary

Acute infection: a productive (lytic) viral infection.

Adjuvant: a formulation, co-administered with an antigen, that potentiates the immune response to that antigen.

Ambisense: of a single-stranded nucleic acid molecule, containing both negative sense and positive sense regions, sometimes written as '$\pm$ sense'.

Antigen: a protein or other molecule which can stimulate an immune response or react with elements of that response. One antigen may contain many epitopes.

Apoptosis: programed cell 'suicide', may be due to cellular damage or in response to external stimuli.

Arbovirus: arthropod-borne virus. A functional designation for viruses transmitted by insect and other arthropod vectors.

Attenuation: decreasing the ability of a virus to cause disease.

B cells: circulating lymphocyte cells that are responsible for serological immunity. They are stimulated to produce antibody by the presence of specific antigens.

Bacteriophage: a virus that infects bacteria.

Baculovirus: a virus infecting insects, a member of the family *Baculoviridae*.

Blunt-end ligation: joining a nucleic acid to another without the overlapping 'sticky end' sequences generated by restriction endonucleases.

Budding: envelopment of a viral nucleocapsid in cell membrane during passage through it.

Cap: a 5′–5′ linked methylguanylate group not coded for by the nucleic acid template and located at the 5′ end of almost all eukaryotic mRNAs, which is normally required for the initiation of translation.

Capsid: the protein component of the nucleocapsid.

Capsomer: a proteinaceous subunit of a capsid.

cDNA: complementary DNA. A DNA copied from an RNA molecule.

Cell-mediated immunity: see *T cells*.

Chemokines: small cytokines controlling the migration and activation of immune cells.

Clonal selection: proliferation of one type of cell recognizing a particular antigen in response to stimulation with that antigen. A basic feature of the immune response.

Concatamer: multiple virus genomes joined together in a polymeric chain.

Cytokine: a protein produced by cells (rather than specialized organs) that controls other cells. A common effector mechanism in the immune system.

Dodecahedron: a 12-sided solid with faces formed of identical pentagons.

Electrophoresis: resolution of biological macromolecules separated by size, resolved by movement under the influence of an electrical current through a gel matrix.

Emerging disease: a disease of infectious origin with an incidence that has increased within the last two decades, or threatens to increase in the near future.

Endemic: established within a geographical area, with ongoing transmission.

Endogenous retrovirus: a viral element present in cellular DNA which derives from an integrated retroviral provirus. May be incomplete.

Envelope: a membrane, derived from the host cell but with viral proteins embedded in it, that surrounds the capsid.

Epidemic: a large-scale outbreak of a disease.

Episome: a plasmid that replicates in eukaryotic cells. Some viral genomes are maintained as episomes during low level infections.

Epitope: a region of an antigen which stimulates an immune response. The structure of individual epitopes can vary widely.

Family: a taxonomical classification, above subfamily, ending in '-*viridae*'.

Fusogenic: able to mediate membrane fusion.

Gene therapy: the introduction of a therapeutic gene into cells in order to correct some undesirable property of those cells.

Genome: the nucleic acid genetic material. In a virus this may be DNA or RNA, but not both.

Genus: a taxonomical classification, below subfamily, ending in '-virus'.

Glycoprotein: a protein with covalently attached sugar groups.

Helical symmetry: a geometric structure of capsid subunits where these are arranged in a rod-like structure with protein subunits in a helix around the nucleic acid.

Helper virus: a replication-competent virus that provides functions necessary for the replication of a defective virus.

Hemorrhagic fever: functional designation of a group of (viral) diseases in which infection causes widespread hemorrhaging at sites throughout the body, primarily by blocking and permeabilizing capillaries. Characterized by sudden onset and severe effects.

Hybridization: reaction of a 'probe' nucleic acid (usually labeled in some way so that it can be easily detected) with other nucleic acids in order to determine whether a nucleic acid with a complementary sequence is present in the test material.

Icosahedral symmetry: a geometric structure of capsid subunits where these proteins are arranged in a roughly spherical shape with two-, three- and fivefold rotational axes of symmetry. The actual shape of the capsid need not be icosahedral and can be some other, broadly similar, shape, notably spherical or dodecahedral.

Icosahedron: a 20-sided solid with faces formed of identical equilateral triangles.

Immunogenic: capable of stimulating an immune response.

Immunoprophylaxis: prevention of infection and/or disease by stimulation of the immune system.

Immunotherapy: moderation of pre-existing disease by stimulation of the immune system.

Introns: sequences removed from an RNA transcription product before it is used as an mRNA.

Latent infection: an inactive state where only part of the viral genome is active.

Ligation: joining of one nucleic acid with another, using 'ligase' enzymes.

Liposome: a spherical structure formed of lipid, usually with a stabilizing agent, which may be fusogenic and can be used in gene therapy or as an adjuvant.

MHC-I: cell surface proteins that present peptide antigens to $CD8^+$ (cytotoxic) T cells.

MHC-II: cell surface proteins that present peptide antigens to $CD4^+$ (predominantly helper) T cells.

Monoclonal antibody: antibody of a single type, specific for a single epitope.

Monospecific antisera: antisera specific for a single antigen, usually containing multiple types of antibody.

mRNA: messenger RNA. An RNA used for translation.

Multivalency: the formulation of a single vaccine to induce immunity to multiple viruses.

Negative sense: of a single-stranded nucleic acid molecule, containing the complementary (opposite) base sequence to the mRNA produced from it.

Nucleocapsid: the viral core, containing structural and replicative proteins together with the nucleic acid genome.

Nucleoside analog: a compound, often an antiviral drug, which is similar to a nucleoside (one of the molecular subunits which make up nucleic acids).

Oncogene: a gene associated with cellular transformation or oncogenesis.

Order: a taxonomical classification above Family, ending in '-*virales*'. Only a few orders have been defined, of which two contain viruses that infect humans.

Pandemic: a world-wide epidemic.

Plasmid: a covalently closed circular DNA molecule able to replicate itself within a host cell.

Plasmid vector: a covalently closed circular DNA molecule designed to express a gene inserted into a specific site within it, derived from a plasmid.

Polyadenylation: a modification found on almost all eukaryotic mRNAs, whereby a tract of adenylate residues (not coded for by the nucleic acid template) is added to the 3′ end of the molecule.

Polymerase: an enzyme that produces nucleic acids by transcription from a complementary template.

Polymerase chain reaction: a DNA amplification system by which a short target DNA is amplified 10^9–10^{12} times for diagnosis, cloning or other applications. Related systems allow amplification of RNA.

Polypeptide: a long chain of amino acids from which a protein is derived. This usually involves a highly variable degree of post-translational modification.

Positive sense: of a single-stranded nucleic acid molecule, containing the same base sequence as the mRNA produced from it.

Prion: a small proteinaceous infectious particle which resists inactivation by agents which destroy nucleic acid and contains an essential modified isoform of a cellular protein.

Prion protein: a cellular surface protein that is present in abnormal form in transmissible spongiform encephalopathies. An aberrantly folded form of this protein is thought to be the infectious agent in these diseases.

Prodrug: a chemical form in which a drug is administered, but which undergoes metabolic conversion to an active form within the body.

Proteinase: an enzyme that cuts a protein within the molecule, usually at a specific amino acid sequence.

Provirus: a DNA copy of the (RNA) retrovirus genome.

Receptor: macromolecular structures that mediate virus–cell interactions.

Representational difference analysis: a technique for isolating DNA present only within certain cells by reacting them with DNA from other cells from the same individual.

Restriction endonuclease: an enzyme that cuts DNA within the molecule at a specific base sequence, usually of four to six bases.

Retrovirus: a member of the class *Retroviridae*; viruses that copy their RNA genome into DNA using a viral reverse transcriptase enzyme and integrate it into the cellular DNA as an essential part of their life cycle.

Reverse transcriptase: an enzyme originally found in retroviruses that mediates 'reverse transcription', producing DNA from RNA.

Ribozyme: an RNA molecule that can catalyze the cleavage of itself, of another RNA molecule, or other reactions.

Scaffolding protein: a protein required for the assembly of a structure but not present in the mature form.

Segmented genome: a genome made up of multiple different segments of nucleic acid, all of which are required to form the complete genome.

Serological immunity: see *B cells*.

Serotype: a grouping of related viruses within a viral species defined by their similar antibody binding characteristics.

Southern blotting: a hybridization system whereby nucleic acids resolved by electrophoresis are transferred to a membrane and detected with specific nucleic acid 'probes'.

Subfamily: a taxonomical classification, above genus and below family, ending in '*-virinae*'. Not all virus families have subfamilies.

Symmetry: the geometric arrangement of protein subunits that make up a capsid. Not necessarily the same as the shape of the capsid. See *Helical symmetry* and *Icosahedral symmetry*.

T cells: circulating lymphocyte cells that are responsible for cell-mediated immunity. Include cells involved in cell killing (cytotoxic T cells), and in immune control and stimulation (helper T cells). Also includes cells involved in delayed-type hypersensitivity, immunological memory and limitation of the immune response.

Template: a nucleic acid used for transcription or reverse transcription.

Temporal control: the division of viral macromolecules into groups synthesized at different times after infection.

T_H1: subset of helper T cells that stimulate the cytotoxic (cellular) response.

T_H2: subset of helper T cells that stimulate the humoral (antibody) response.

Transcription: the production of RNA with a complementary sequence (A for T, C for G, G for C, U for A) from DNA.

Translation: the production of protein from mRNA.

Transmissible spongiform encephalopathy: one of a class of neurodegenerative diseases characterized by the formation of lesions within the brain associated with abnormal prion protein.

Vacuole: a membrane-bound cellular vesicle of which many subtypes exist. These are often involved in intracellular transport or in the digestion of macromolecules taken up by the cell.

Vector: in cloning, a modified plasmid used to carry, replicate and (possibly) express inserted DNA. Of a vaccine, either a replication-competent nucleic acid into which a gene coding for an immunogen is inserted (live or gene vector), or a carrier protein to which a (smaller, immunogenic) protein or peptide is attached (fusion vector). In epidemiology, an agent responsible for spread of a disease, often an insect or rodent.

-virales: suffix denoting an order of viruses.

-viridae: suffix denoting a family of viruses.

-virinae: suffix denoting a subfamily of viruses.

Virino: an alternative (hypothetical) infectious particle to the prion, containing a small nucleic acid and a modified isoform of a cellular protein.

Virion: a virus particle.

-virus: suffix denoting a genus of viruses.

Western blotting: an immunoassay system whereby proteins resolved by electrophoresis are transferred to a membrane and detected with specific antisera.

Zoonosis: the transfer of an infectious agent from an animal source.

Index

Abortive infection, 29
Aciclovir, 39, 100–104, 106, 108, 111–112
  resistance, 106–107
  triphosphate, 100–103, 112
Acquired immune deficiency syndrome, 5, 69, 97, 104, 151, 156, 158–159, 164, 169
Acute infection (*see* Lytic infection)
    respiratory distress syndrome, 154
Acyclic sugar, 100, 102
Acyclovir (*see* Aciclovir)
Additive drug effects, 107–108
Adenosine
  arabinoside (*see* Vidarabine)
  deaminase deficiency, 131
Adenoviridae, 3–4, 15–18, 43, 68–69
Adenovirus, 7, 9–14, 18–19, 36, 44, 92–93, 96, 128, 131–132
Adeno-associated virus, 18, 45, 92, 131–132
Adjuvant 57, 73–74, 84, 86–90, 95, 173, 175
AIDS (*see* Acquired immune deficiency syndrome)
Alum, 87–89
Amantadine, 101
Ambisense, 4–5, 27, 33, 173
Amherst, Lord Jeffrey, 160–161
Ampicillin, 122–123, 126
Amplification of
  disease, 159, 166, 168–169
  nucleic acid, 76
Antagonistic drug effects, 107–108
Antibiotics, 47, 99–100, 104, 107, 112, 123, 126
  resistance, 157, 161
Antibody, 55, 60–66, 73, 75, 77, 82, 88, 90–91, 97, 140, 146, 148, 173, 175
  affinity, 66
  anti-idiotypic, 90–91
  dependent cellular cytotoxicity, 55, 61, 64–65
  humanized, 78
  monoclonal, 63, 73–78, 175
  monospecific (*see* Monospecific antisera)
  neutralizing, 95
  non-neutralizing, 94
  recombinant, 75–78
  therapeutic, 66, 72, 76
Antigen, 55–62, 64, 66, 71, 73–74, 77, 85–87, 89–91, 95, 135, 137, 146, 173, 175
  presenting cells, 56–59, 61, 67–68, 87
Antigene, 116
Antigenic
  drift, 27, 69–70, 94, 152, 156–157
  shift, 27, 69–70, 152–158
Antisense, 115–117
Antisera, 72, 136, 148
  monospecific, 72, 175
Antiviral
  drugs, 39, 78, 99–119, 152, 157
    combinations (*see* Combination therapy)
    development, 110–118
    nucleic acid-based, 115–118, 129
    rational design, 113–115
    resistance, 104–105, 106–109, 152, 157
    resistance, multiple, 107
    toxicity, 99, 104, 109, 112–113, 117
  immunoglobulin, 76–78
  state, 53
Apoptosis, 49, 55, 58, 62, 67–69, 95, 127, 173
Ara–A (*see* Vidarabine)
Arabidopsis, 10
Arbovirus, 154, 173
Arenaviridae, 3–4, 17, 33, 106, 153
Arenavirus, 8
Arteriviridae, 6
Arthropod, 173
Ascites, 75
Assembly of viruses, 2, 17, 23–24, 29–31, 34, 36–37, 101, 106, 127
Astroviridae, 3–4
Attentuation, 173 (*see also* Vaccine, live attenuated)
  defined, 84–86, 90, 95
  random, 84–85
Autoimmunity, 70
Avipoxvirus, 10

Axes of symmetry, 9, 11
Azidothymidine, 100–104, 107–108
  triphosphate, 103–104
AZT (*see* Azidothymidine)

β-galactosidase, 123
β_2 microglobulin, 58, 70
B7 56, 61, 68
B19 infection, 5 (*see* also Parvovirus)
B cell, 56, 59–64, 68, 75, 173
  epitope, 63, 70–72
  memory, 64
  receptor, 64–65
Bacilllus megaterium 10
Bacteriophage, 19, 36, 41, 75–77, 161, 173
  filamentous, 75–77, 122–123, 126, 127
  lambda, 123, 127
  M13, 127
  T7
    polymerase, 145
    promoter, 126
  therapy, 161
Baculoviridae, 173
Baculovirus, 124–125, 127–129, 152, 162–163, 173
bcl-2, 67, 69
Biocatalytic mechanism, 48–49
Biological
  control, 162–163
  weapons, 152, 160–161
Birnaviridae, 15, 17
Block co-polymer, 88–89, 96
Blood, 136, 148, 154, 168
Bornaviridae, 6
Bovine
  growth hormone polyadenylation site, 126–127
  spongiform encephalopathy, 49–50, 152, 156, 166–167
Branched DNA assay, 140
Brovavir, 110–111
BSE (*see* Bovine spongiform encephalopathy)
Budding, 32, 35–38, 173
Bunyaviridae, 3–4, 17, 33, 153, 158
Burkitt's lymphoma, 43–44
BvaraU (*see* Brovavir)

882C (*see* 5-propynyl-ara-U)
Caenorhabditis elegans, 10
Caliciviridae, 3–4, 15, 17, 162
Calicivirus, 156, 160
Canarypox, 92
Cancer, 22, 33, 41–43, 97, 100, 109, 111, 133, 163, 165, 168
Canyon, 115
Cap, 28–32, 173
Capsid, 3–4, 8, 11–12, 21, 31–32, 36, 173–174
Capsomer, 8, 173
Capture tag, 144, 147
Carcinoma, 25, 43
Carrier protein, 73
Caulimoviridae, 24
CC-CKR5, 18
CD
  2, 58–59
  3, 58
  4, 16, 34, 56, 59, 61, 69 (*see also* T cell, $CD4^+$)
    soluble, 101
  8, 56 (*see also* T cell, $CD8^+$)
  21, 18
  28, 56, 61, 68
  40, 60–61, 64, 68
  40 ligand, 60–61, 68
  46, 18
  55, 18
  58 (*see* LFA-3)
cDNA, 85, 123–124, 127, 139, 173
Cell
  culture, 39–40, 53, 75, 84, 124–125, 135
  cycle, 44
  differentiation, 131
  killing, 40, 127–128, 132
  lysis, 28, 31, 37, 39
  -mediated immunity (*see* Immune response, cell-mediated)
  tropism, 131
Cerebrospinal fluid (*see* CSF)
Cervical carcinoma, 43
Chain terminator, 105
Chemokine, 173
  receptor, 34
Chickenpox (*see* Varicella)
Chronic infection (*see* Persistent infection)
Cidofovir, 105
Circoviridae, 15
Circovirus, 10, 22
cis-activation/repression, 43
CJD (*see* Creutzfeld-Jakob disease)
Class switching, 64
Clinical trials, 93–95, 117, 133
Clonal selection, 60, 62, 173
Cloning, 71, 82, 85–86, 90–91, 109, 121–129, 135, 139, 164, 177
  site, 123, 125–126
  vector
    bacteriophage, 128
    viral, 127–129, 162
Co-factor, 44
Co-stimulation, 56, 59, 61–62, 67–68
Codon, 13–14
Colorado tick fever, 5, 153
Combination therapy, 104, 107–109, 156
Common cold, 4
Complement, 53, 55, 64, 68–70
  activation, 65
  fixation, 136–137
Complex morphology, 3–6
Concatamers, 22, 174
Coronaviridae, 3–4, 6, 15, 17–18
Coronavirus, 18
Cosmid, 123, 128

Cowpox, 81, 162
Coxsackievirus, 18
Creutzfeld-Jakob disease, 48–50, 167
  new variant, 49–50, 167
Crixivan (*see* Idinavir)
Crystallography, 71–72
CSF, 136–137
CTL (*see* T cell, cytotoxic)
CTLA-4, 68
CXCR4 (*see* Fusin)
Cyclin, 44
  -dependent kinase, 42
Cymevene, 105
Cystic fibrosis, 129, 131
Cytokines, 39, 53–54, 56, 58–60, 64, 67–69, 78, 87, 89, 95, 100, 173–174
Cytomegalovirus, 18, 46, 68–70, 107, 110, 112, 141, 145
  immediate-early promoter, 126
Cytoplasm, 18, 20, 22, 28, 36, 45, 55–58, 93
Cytoplasmic pathway of antigen presentation, 57, 89, 91
Cytotoxic T cell (*see* T cell, cytotoxic)

D4T (*see* Stavudine)
DDC (*see* Zalcitabine)
DDI (*see* Didanosine)
Decoy oligonucleotide, 116–117
Defective
  interfering virus, 40, 47
  virus, 44–47, 121, 174
Delayed-type hypersensitivity, 60, 176
Delta agent (*see* Hepatitis D virus)
Delta antigen, 46
Dengue
  fever, 152–155, 168
  hemorrhagic fever, 154–155, 158
Depot effect, 87
Detection systems, 139, 144–147
Diagnosis, 135–149, 152
Diagnostic kits, 135, 146, 148
Diarrhea, 4, 81
Didanosine, 105
Dideoxyinosine (*see* Didanosine)
Differentiation, 60
DISC vaccine, 90, 93, 97
Dodecahedron, 11, 173
Downer cow syndrome, 49, 167
Drowning sickness, 154
Drug abuse, 168–169

E1A protein, 44
E1B protein, 44
E6 protein, 44
E7 protein, 44
EBNA, 44
Ebola, 8, 76, 152, 166, 168
  Reston, 159, 166
Echovirus, 18
Eclipse phase, 19–20
Eggs, 73, 109, 154
Electron microscopy, 3, 6–8, 10, 136
  immune, 135–136
Electronic information resources, 52, 79, 98, 119, 133, 149, 171
Electrophoresis, 136–138, 141, 144–146, 174, 177
Elimination of diseases, 81, 85, 162
ELISA, 136–137, 145
Emerging
  diseases, 151–170, 173 (*see also* New viruses)
    definition, 151
    factors controlling, 167–169
    monitoring, 169–170
Emulsion, 88–89, 96
Encapsidation, 24, 28, 33
Encephalitis, 5
Endemic, 153, 174
Endocytosis, 16, 29
Endogenous
  retroviruses, 35, 41, 47, 166, 174
  viral sequences, 47
  viruses, 85
Endoplasmic reticulum 30, 57–58
Endosomal pathway of antigen presentation, 56–57, 59
Endosome, 16, 19, 28, 32
Enteric coating, 96
Enterovirus, 110, 141
Entry of viruses, 19
Envelope, 8–9, 21, 32, 174
Enveloped, 3–7, 19, 36
Enzyme, 12, 20, 22, 27, 30, 37, 39, 45–46, 68, 99, 102, 104, 107, 112, 115–116
  restriction (*see* Restriction endonuclease)
Epidemic, 151, 174–175
Epidemiology, 33, 177
Episomes, 21, 33, 39–40, 44, 47, 121, 128, 131–132, 174
Epitope, 69–73, 82, 89–91, 173–174 (*see also* B cell, epitope; T cell, epitope)
  conformational, 63, 70–72
  discontinuous, 63, 70–72
  disperse, 63, 70–72
  linear, 63, 70–72
  polysaccharide, 91
  post-translational, 70–72
Epivir, 105
Epstein–Barr virus, 13, 18, 43–44, 69, 86, 94, 110, 128, 165
Equine morbillivirus, 153
Erythrocyte P globoside antigen, 18, 23
Erythrocyte precursor cells, 23
Escherichia coli, 10, 127, 140, 143
Eukaryotic
  expression systems, 86
Exanthem, 5

f1 ori, 122, 126
$F_{a,b}$ region, 63

F_c
  receptor, 64, 70
  region, 63
Famciclovir, 100, 105, 111–112
Family, 4, 6, 174, 176–177
Famvir (*see* Famciclovir)
Fatal familial insomnia, 48
Feces, 153–154
Fibrils, 48–49
Filoviridae, 3–4, 6, 16–17
Filovirus, 8–9
Flaviviridae, 3–4, 15, 81, 106, 148, 153–154, 164
Flavivirus, 44
Fleas, 160
5-Fluorouracil, 111
Foot-and-mouth disease virus, 18
Formalin, 94
Foscarnet, 105, 107–108
Four Corners, 153–154
  virus (*see* Sin Nombre virus)
Fox, 96, 163
Frameshifting, 14–15
Freund's adjuvant, 87–88
Fungi, 10
Fusin, 18
Fusion, 174–175

γ/δ cells, 55
G418, 126
Gammaherpesvirinae, 165
Ganciclovir, 105, 112
Gastric cancer, 43
Gastroenteritis, 4
Gene gun, 93
Gene therapy, 117–118, 128–133, 163, 174–175
  endogenous, 129–131
  exogenous, 129–131
  germline, 130, 132
Genetic
  susceptibility to disease, 50
  vaccination (*see* Vaccine, nucleic acid)
Genome, 1–2, 4–17, 19–20, 22, 24, 28, 30, 32–34, 36, 40–42, 44–47, 73, 84–85, 92, 115–118, 123–124, 128, 142, 152, 157, 164, 174–176
  library, 128
  replication, 21–22, 24–25, 31, 33
  segmented, 26, 32–33, 69–70, 152, 157, 176
  size, 10–15, 93, 25
  type, 15–16
Genus, 6, 174, 176–177
Gerstmann–Straussler syndrome, 48, 50
Glycoproteins, 2, 10, 16, 32, 35–36, 62, 69, 71, 93, 94, 157–158, 163, 174 (*see also* Protein, glycosylation)
Granzymes, 55, 58
Guanosine, 100, 102–103

Hantaan, 153–154, 170
Hantavirus, 152–154, 156, 158, 170
  pulmonary syndrome, 153–154
Helical
  capsid, 3–8, 11, 32
  symmetry, 9, 174, 176
Helper virus, 34, 45–46, 121, 127, 174
Hemagglutinin, 157
Hematopoietic stem cells, 132
Hemophilia, 129
Hemorrhagic fever, 4, 8, 151, 154–155, 169, 174
  with renal syndrome, 153–154
Hepadnaviridae, 3–4, 16–18, 24–25, 43, 106
Hepadnavirus, 14, 44, 136
Heparan sulfate, 18
  proteoglycan, 18, 21
Hepatitis, 58
  A, 4
    vaccine, 83
    virus, 18, 148
  B, 4, 109, 169
    surface antigen, 86
    vaccine, cloned, 83, 90
    vaccine, subunit, 83
    virus, 18, 23, 24–25, 43–44, 95, 101, 106, 112, 141, 148, 156
  C, 109, 135, 152, 157, 169
    virus, 44, 100–101, 106, 141, 164–165
    virus discovery, 148
  D 'virus', 46
  E virus, 3
  non-A non-B, 148, 164
Hepatocellular carcinoma, 25, 43
Herpes
  B virus, 153
  simplex, 4, 93, 96, 100–101, 112
    encephalitis, 55
    virus, 13, 18, 21–22, 37, 41, 69, 100–101, 104–105, 131
Herpesviridae, 3–4, 15, 17–18, 20, 36, 43, 68–69, 105, 153
Herpesvirus, 7–8, 13–14, 19–20, 22, 39, 44–45, 66, 68–70, 86, 92–93, 95–96, 100, 105, 107–108, 110–112, 115, 121, 128, 131–132, 138
  saimirii, 165
HIV (*see* Human immunodeficiency virus)
Hivid (*see* Zalcitabine)
Host shutoff, 37
HPMPC (*see* Cidofovir)
HTLV (*see* Human T-cell leukemia virus)
Human genome project, 129
Human herpesvirus
  6, 167
  7, 16, 18
  8, 43–44, 69, 148
    discovery, 165
Human immunodeficiency virus, 16, 18, 26, 34–35, 69–71, 81, 86, 91, 94, 96, 100–110, 113–114, 117, 129, 141, 145, 152, 165, 168–169
  gag polyprotein, 35, 104
  gp120, 34–35, 71, 91

rev, 35, 117
tat, 35, 117
V3 loop, 71, 91
Human T-cell leukemia, 5
virus, 43–44, 141
HVEM, 18, 21
Hybridization, 123, 139–141, 145, 174, 176
dot-blot, 137, 139–140
in situ, 136–137, 139–140
microplate, 145, 147
Hybridoma, 73–75
3′-hydroxyl group, 100
Hyperexpression, 129

ICAM
-1, 18, 58–59
-2, 58–59
Icosahedral
capsid, 3–8, 11–12, 36
symmetry, 9, 174, 176
Icosahedron, 9, 174
Identification of viruses, 148, 153, 164–165, 167
Idoxuridine, 105
Ig (*see* Immunoglobulin)
Immediate-early, 21–22
Immortalization of cells, 42
Immune
evasion by viruses, 68–70
memory, 56, 61
response, 57, 75, 78, 93–94, 136, 158, 173–176
antibody (*see* Immune response, serological)
cell-mediated, 55–60, 66, 84, 88–91, 93, 95, 173, 177
humoral (*see* Immune response, serological)
mucosal, 96–97
non-specific, 53–55, 93
pre-specific, 53, 55
protective, 84, 86–87, 89–90, 93–94
serological, 60–68, 91, 93, 95, 173, 176–177
tailoring, 89, 95–97
tolerance, 40, 59, 67
Immunization, 73–75, 87–88
Immunoassay, 72, 135–137, 140
Immunocytochemistry, 135–136
Immunodominance, 66
Immunofluorescence, 135–137
Immunogenic, 71, 84, 86–87, 91–92, 94, 125, 175
Immunoglobulin, 60–61, 70
A, 64–65, 169
receptor, 18, 24
secretory, 65–66, 97
chain types, 65
complementarity-determining regions, 61, 78
constant region, 61, 64, 66
D, 62–65, 94
E, 63–65, 169
G, 64–66, 135
structure, 60, 63
subtypes, 64
hinge region, 63
isotype, 64
M, 62–66, 135–136
variable region, 61, 76
'Y', 73
Immunological memory, 176
Immunomodulator, 87, 89, 95, 109
Immunoprophylaxis, 175
Immunosuppression, 4, 92, 107, 131, 166, 168–169
Immunotherapy, 175, 163
In situ hybridization (*see* Hybridization, *in situ*)
Indicator gene, 122–123, 128
Inflammation, 64, 68–69, 92
Influenza, 5, 8, 46, 69–70, 85–86, 93, 101, 109, 152, 156–159, 166
A virus, 18, 26, 71, 106
H5N1, 158
Hsw1N1, 158, 164
B virus, 18, 26
C virus, 18, 26
vaccine, 83
virus, 19, 28, 32, 54
Inoculation, 84, 96
Insect, 125, 128, 162–163, 173, 177
Insertional mutagenesis, 33, 42, 44
Integrase, 33–35
Integration, 33, 39–42, 44–45, 128, 130–132, 176
Integrins, 18, 67
Interferon, 39, 53–56, 68–69, 100–101, 108–109, 116
alpha, 54–55, 106, 108–109
beta, 54–55
gamma, 54–55, 59
Interleukin, 54, 56
2, 59–60
4, 59–60
5, 59–60
6, 59–60
10, 60, 69
12, 55, 59–60
Internal ribosomal entry site, 29
International Committee for the Taxonomy of Viruses, 2–3
Introns, 15, 127–129, 175
Invirase (*see* Saquinavir)
Iridoviridae, 17
ISCOMs, 89
Isotype (*see* Immunoglobulin, isotype)

J chain, 64
Japanese encephalitis vaccine, 83
JC virus, 41
Jenner, Edward, 81, 162
Jesty, Benjamin, 81, 162

Kaposi's sarcoma, 43, 148, 165
virus (*see* Human herpesvirus 8)

Keratinocytes, 59
Keyhole limpet hemocyanin, 89, 91
Kinase, 54, 68, 101–103 (*see also* Thymidine kinase)
Kuru, 48

Labeling (*see* Detection systems)
Lamivudine, 105
Lassa fever, 4, 153
  virus, 106
Late, 21
Latency–associated transcripts, 41
Latent infection, 14, 21, 37, 39–41, 69–70, 86, 92, 118, 128, 140, 175 (*see also* Reactivation)
Leukemia, 43
LFA
  -1, 58–59
  -3, 58–59
Licensing, 92, 113
Ligase, 174
  chain reaction, 144–145
Ligation, 124, 144–145, 173, 175
Lipid A, 87, 89
Liposome, 57, 89, 96, 130, 175
Logarithmic growth phase, 20
Long terminal repeat, 35
Lung, 129, 132
Lymphocytes, 131–132 (*see also* B cell, Peripheral blood mononuclear cells, T cell)
Lymphocytic choriomeningitis virus, 58, 94
Lymphoma, 43
Lysogeny, 41, 47
Lysosome, 28
Lytic infection, 37, 39–40, 173

Macrophage, 55
Mad cow disease (*see* Bovine spongiform encephalopathy)
Mannose-6-phosphate receptor, 18, 21
Marburg, 8, 152, 166
Marker gene (*see* Indicator gene)
Measles, 5, 41, 81, 152, 158
  vaccine, 83
  virus, 18, 40, 70
Membrane, 16, 18, 22, 32, 34, 36–37, 61, 64, 67, 173–174, 177
Metallothionein, 126
Methylguanylate (*see* Cap)
MHC, 67
  I, 55–58, 68–71, 87–89, 91, 175
  II, 55–61, 69, 71, 88, 175
  restriction, 58
Mimicry, 69–70
Mink, 48–49
MMR vaccine, 91
Mobile genetic elements, 45, 47
Molluscum contagiosum, 5
Monkey, 152–154, 156, 159, 165–166
  cells, 83, 85
Monkeypox, 163–164
Monoclonal antibodies (*see* Antibody, monoclonal)
Mononegavirales, 6
Monospecific antisera (*see* Antisera, monospecific)
Morphology, 3–8, 136
Mosquito, 153–155, 159–160, 168
Mouse, 73–74
  Asian striped field, 153
  deer, 154
mRNA, 2, 13–15, 17, 21, 23–25, 28–31, 33–34, 37, 41, 75, 100, 115–116, 118, 123–124, 127, 137, 174–177
MRSA, 157
Mucosal surfaces, 65, 93, 96–97
Multivalent (*see* Vaccine, multivalent)
Mumps, 5
  vaccine, 83
Muramyl
  dipeptide, 87–89
  tripeptide, 89
Muscle, 93
Mutagenesis, 122, 126, 128
Mutation, 26–27, 35, 44, 66, 69, 70, 77, 82, 84–85, 106–108, 152, 156–157
  point, 85
Mycobacterium tuberculosis, 157
Mycoplasma genitalium, 10
Myeloma, 43, 73–74, 77
Myxoma, 43, 160
Myxomatosis, 152, 155, 160
Myxoviruses, 106

NASBA, 137, 144–145
  quantitative, 145
Nasopharyngeal carcinoma, 43–44
Natural killer cells, 53–56, 59, 64, 67, 69
Negative sense, 4–5, 15, 17, 25, 27–31, 33, 175
Nelfinavir, 105
Neomycin, 126
Neuraminic acid, 18, 28, 32
Neuraminidase, 157–158
Neurone, 128, 131, 167
Neurovirulence, 86
Neutralization, 63–64
Nevirapine, 100, 105, 107
New viruses (*see* Emerging diseases)
Nidovirales, 6
NK cells (*see* Natural killer cells)
Non-enveloped, 3–8, 19, 66
Non-nucleoside reverse transcriptase inhibitors, 100, 104–105
Norvir (*see* Ritonavir)
Norwalk, 4
Nucleic acid
  amplification, 135, 137, 139–147 (*see also* Ligase chain reaction, NASBA, Polymerase chain reaction)
  detection, 135–140

drugs (*see* Antiviral drugs, nucleic acid-based)
hybridization (*see* Hybridization)
probes, 116, 136, 139, 145–147, 174, 176
sequence, 73, 83, 137, 139, 146–148, 164, 169
vaccination (*see* Vaccine, nucleic acid)
Nucleocapsid, 2, 8–9, 18, 24–25, 31–32, 34, 36, 173, 175
Nucleoside, 102
analog, 99, 101–102, 104–112, 175
Nucleus, 22, 34, 36, 46, 67
nvCJD (*see* Creutzfeld-Jakob disease, new variant)

2′-5′ Oligo_A synthetase, 54
Oligonucleotide, 116, 139, 141
Oncogenes, 33, 42–43, 47, 67, 175
Oncogenesis, 39–44, 47, 84, 92, 129, 175, 132
One-step growth curve, 19–20
Open reading frame, 25, 46, 91–92
Oral
bioavailability of drugs, 111–112
polio vaccine (*see* Polio, vaccine, live oral)
Order, 6, 175, 177
Origin of replication, 122–123, 125–126
Orphan disease, 164
Orthomyxoviridae, 3, 5, 16–18, 26, 31, 36, 69, 106
Overlapping genes, 14

p53, 42, 44, 67
Pandemic, 85, 151, 156–158, 175
Papilloma, 5, 43
Papillomavirus, 18, 43, 86, 96, 100, 106, 131, 136, 141
high risk, 43–44
Papovaviridae, 3, 5, 15, 17–18, 22, 43, 85, 106
Papovavirus, 13, 22, 165
Paramyxoviridae, 3, 5–6, 16, 17–18, 31–32, 36, 69, 106
Paramyxovirus, 7, 19
Parvoviridae, 3, 5–6, 15, 17–18, 22, 45, 69, 131
Parvovirus, 12–14, 22–23, 45, 92, 132
Passive immunity, 66, 76–78
Patent, 140
Penciclovir, 111–112
triphosphate, 112
Pepscan mapping, 71–72
Peptide, 56, 62, 71, 90, 92, 175
conjugation, 73, 90, 177
MHC-I presentation of, 57–58, 69, 89, 91
MHC-II presentation of, 58–60, 69
mimetic, 105, 106
synthetic, 71–73, 82
Peripheral blood mononuclear cells, 78
biased, 75–76
naïve, 75–76
Perforin, 55, 58
Peroxidase, 140
Persistent infection, 39–40, 47, 58
pH, 16, 28, 32
Phagocytes, 64, 67
Phosphorothioate, 116
Phosphorylation, 101, 104, 106 (*see also* Protein phosphorylation)
Picornaviridae, 3, 5, 15–18, 29, 69
Picornavirus, 11, 13–14, 17, 101, 115
Pig, 157, 166
Plants, 10, 45
Plasma cells, 60–62, 64, 66
Plasmid, 47, 90, 93–94, 121–123, 125, 128, 139, 157, 174–175, 177
vector, 121–125, 175
Plasmodium falciparum, 10
Polio, 5, 81
vaccine
contamination of, 165
genetic stabilization of, 85
inactivated, 83, 85
live oral, 83, 85, 91, 96
virus, 12, 18–19, 29–30, 36, 66, 83, 86, 92–93
wild, 83
Polyadenylation, 29, 175
site, 126–127
Polycloning site, 122, 126
Polyhedrin, 129, 162
promoter, 127
Polymerase, 22, 23–24, 26–33, 46, 102–103, 106–107, 122, 140, 144–146, 175
chain reaction, 76, 116, 123–124, 137, 140–147, 175
avoiding contamination, 146–147
in situ PCR, 142
multiplex, 142
nested, 142
quantitative, 142, 146
RT–PCR, 137, 142
thermostable, 140, 141–143, 147
Polyomavirus, 41, 43–44, 125
Polypeptide, 175
Polyprotein, 14–15, 29–30
Population movements, 158–159, 168
Positive sense, 4–5, 15, 17, 25, 27–30, 33, 176
Post-translational modifications (*see* Protein, post-translational modifications)
Potato spindle tuber viroid, 10
Potentiation of disease, 94
Poxviridae, 3, 5–6, 15, 17–18, 20, 43, 68–69
Poxvirus, 7–9, 12–13, 91–93, 96, 99, 121, 160, 163–164
Primers, 22, 31, 116, 141–142, 144–145, 147
Prion, 45, 47–49, 156, 176–177
definition, 48
Prodrug, 100–101, 112, 176
Progressive multifocal leukoencephalopathy, 5
Prokaryotic cloning systems, 86, 122–129
Promoter, 21, 31, 93, 121–122, 124–126, 128
conditional, 117, 126, 131
Proof-reading, 26, 143, 147

Prophylaxis, 97–98
5-Propynyl-ara-U, 110
Protease (*see* Proteinase)
Proteasome, 57
Protection (*see* Immune response, protective)
Protein
  acylation, 125
  folding, 125
  function, 125
  glycosylation, 21, 32, 35, 63, 65, 70, 72, 90, 124–125, 128
  phosphorylation, 37, 124–125
  post-translational modifications, 70, 72, 90, 101, 123–125, 128–129, 175
  secretion, 125
Proteinase, 30, 34–35, 42, 48, 57–58, 114–115, 142, 176
  aspartyl, 104, 114
  inhibitors, 100–101, 104, 108–109, 113–114
  serine, 104
Proteolysis, 72, 125
Proton pump, 16
Provirus, 33–34, 39–41, 108, 174, 176
PrP, 49
  ^{C} form, 48–49
  ^{Sc} form, 48–49, 125
  gene, 50, 167
    homozygosity, 50
pUC, 19 122
Purkinje cells, 49
Pyrophosphate, 105

Quarantine, 159

Rabbit, 73, 155, 160–162
  hemorrhagic disease, 152, 156, 160–162
Rabies, 5, 18, 153, 163
  vaccine, 83, 98
Raccoon, 96, 163
Radioactivity, 78, 139
Radioimmune precipitation, 63
Radioimmunoassay, 136–137
Rat, 73–74, 96, 153–154
Rb (*see* Retinoblastoma)
Reactivation, 95, 98, 109, 118
Reading frame, 15
Reappearance, 163–164
Receptors, 176
  cellular for viruses, 1, 16, 18, 131
  immunological, 60, 64
  viral, 101, 115
Recombinant phage antibody system 75–77
Recombination, 35, 76, 152, 157–158
Release of virus
  accidental, 152, 161–162
  deliberate, 152, 160–161
Reoviridae, 3, 5, 15, 17–18, 69, 81, 158
Reovirus, 18, 28, 29, 37
Replication of viruses, 1–2, 16–37, 41, 47, 55
Replicative
  form, 22–23, 31
  intermediate, 23, 29–32
Reporter
  gene, 126
  tag, 139, 144, 147
Representational difference analysis, 164, 176
Reservoir of infection, 26, 154, 166
Respiratory
  disease, 4–5, 109
  syncytial virus, 94, 106
Restriction
  endonuclease, 123–124, 138–139, 173, 176
  fragment length polymorphism, 138–139
Retinoblastoma, 42, 44
Retrotransposons, 47
Retrovir (*see* Azidothymidine)
Retroviridae, 3, 5, 16–18, 33, 43, 69, 81, 105, 176
Retroviruses, 14, 34, 36–38, 41–42, 44, 47, 54, 92–93, 101, 121, 128, 131–132, 166, 176
Revaccination, 84
Reverse transcriptase, 24, 33–35, 47, 103–107, 124, 137, 141–145, 147, 176
  inhibitor, 104–107
Rhabdoviridae, 3, 5–6, 16–18, 31, 153
Rhabdovirus, 19
Rhinovirus, 18, 115
Ribavirin, 106
Ribonuclease
  H, 145
  L, 54
Ribosome, 8
Ribozyme, 46, 116–118, 176
Ricin, 78
Rift valley fever, 4
Rimantadine, 106
Ritonavir, 100, 105
RNase H, 34–35
Ross River Virus, 159
Rotavirus, 18, 81, 85, 138, 158
Rous sarcoma virus, 44
  LTR promoter, 126
Rubella, 5
  vaccine, 83

Saccharomyces cerevisiae, 10
Safety, 94–95, 128–129 (*see also* Clinical trials, Toxicity)
Salamander, 10
Salmonella, 91, 96
Saponin, 88–89
Saquinavir, 100–101, 106, 114
Sarcoma, 43
Satellite virus, 45–46
Scaffolding proteins, 36, 176
Scale-up, 86, 124
Scintillation proximity assay, 146, 147
Scrapie, 48–49, 125, 167
Selectable gene, 122–123, 125–126
Selectivity, 99–100, 104
Self-assembly, 13

Semliki Forest virus, 18
Serological
  evidence of infection, 156
  immune response (*see* Immune response, serological)
Serotype, 83, 91, 155, 158, 176
Serum, 73, 136
Shingles (*see* Zoster)
Shuttle vector, 125
Sialic acid (*see* Neuraminic acid)
Signal transduction, 58, 67
Simian
  immunodeficiency virus, 152
  virus, 40 (*see* SV40)
Sin Nombre virus, 153–154, 156, 170
Skin, 136
Slow release, 96
Slow virus, 41, 48 (*see also* Persistent infection)
Small round viruses, 8
Smallpox, 5, 7, 81, 91, 99, 152, 158, 160, 162–164
  vaccine, 83 (*see also* Cowpox, Vaccinia)
Sorivudine (*see* Brovavir)
Southern blotting, 137, 145–146, 176
Splicing
  DNA, 128
  RNA, 13–14, 61, 127
Squalane, 88
Squalene, 88–89
Stable expression, 121
Stavudine, 106
Sterile immunity, 96–97
Sub-acute sclerosing panencephalitis, 40–41
Subcloning, 125
Subfamily, 6, 173, 176–177
Sub-viral infections, 44–50
Suicide virus, 121
Sulfonamide, 99
Surfactant, 88–89
SV40, 44, 85, 125, 152, 165
  origin of replication, 126
  promoter, 126
Swabs, 136
Swine flu, 157–158, 164, 170
Synergistic drug effects, 107–108

T3, 59
T7 (*see* Bacteriophage T7)
T cell, 55, 67, 176
  CD4+, 56–57, 59, 61, 69, 71, 88, 175
  CD8+, 56–58, 71, 87–89, 91, 175
  cytotoxic, 42, 54, 56–60, 67, 71, 88–89, 91, 175–176
  epitope, 56, 61, 71–72, 89–90
  helper, 56–57, 59–61, 64, 68–69, 71, 175–177
  memory, 55–56
  naïve, 68
  receptor, 54, 56, 58–59, 61
  suppressor, 59

T lymphocyte (*see* T cell)
Tailoring of the immune response (*see* Immune response, tailoring)
TAP transporter protein, 57
Taq polymerase (*see* Polymerase, thermostable)
Taxonomy, 2, 6, 174–176
3TC (*see* Lamivudine)
TCR (*see* T cell, receptor)
Tegument, 8, 21, 37
Temporal regulation, 21, 23, 31, 177
Tetrahymena, 117
T_H1 cells, 60, 95, 177
T_H2 cells, 60, 95, 177
Therapeutic antibody (*see* Antibody, therapeutic)
  gene, 129–132, 174
  vaccination (*see* Vaccine, therapeutic)
Therapy, 96
Thiosemicarbazone, 99
Throat washings, 136
Thymidine, 102
  kinase, 20, 101, 103–104, 106–107, 163
Tick, 5, 153–154
  -borne encephalitis vaccine, 83
Tobacco mosaic virus, 7, 9, 11
Togaviridae, 3, 5, 16–18, 153, 159
Togavirus, 19
Tolerance (*see* Immune tolerance)
Toxicity, 87, 95 (*see also* Antiviral drugs, toxicity)
Toxin, 45, 78, 163
  cholera, 96
  tetanus, 96
trans-activation/repression, 25, 43
Transcription, 13, 21, 25, 31, 33, 42, 177
  control of, 13, 21, 28
Transformation, 39–43, 128, 175
Transgenic animals, 49, 132
Translation, 13, 15, 22, 24–25, 29, 32, 177
  control of, 28, 29–30, 37, 54, 127
Transmissible spongiform encephalopathies, 41, 48–50, 166–167, 176–177
Transposons, 47
Trifluorothymidine, 105
Tumor, 75
  necrosis factor, 54, 56
  suppressor genes, 42–43

Ultraviolet, 48, 147
Uncapped mRNA, 37
Unenveloped (*see* Non-enveloped)
Uracil, 110
  N-glycosylase, 142, 147
Urine, 136, 153–154
URLs (*see* Electronic information resources)

Vaccination, 41, 43–44, 92, 99, 162–163
  mucosal, 96–97
  oral, 96, 128
  therapeutic, 97–98

Vaccine, 63, 66, 71, 73, 81–98, 125, 135, 162–163, 165, 170, 177
anti-idiotype, 82, 91, 95
associated disease, 83
attenuated (*see* Vaccine, live attenuated)
attenuation, 82
cloned, 81–84, 86
development, 90–93
fusion vector, 90–91, 94–95, 177
gene vector, 90, 92–95, 177
inactivated, 71, 81–84, 86, 88
killed (*see* Vaccine, inactivated)
live attenuated, 71, 81–86, 94, 95
live vector, 81–82, 85–86, 90–96, 128–129, 177
multivalent, 85, 91–92, 95, 175
nucleic acid, 81–82, 90, 93–95
peptide, 95
subunit, 81–84, 86, 88–89, 94–95
synthetic, 82
Vaccinia, 7, 18, 91, 96, 128, 152, 163
Vacuole, 18, 22, 37, 39, 177
Valaciclovir, 105, 111–112
Valtrex (*see* Valaciclovir)
Varicella 4, 100, 151
vaccine 83, 92, 97
Varicella-zoster virus 18, 39, 98–100, 105, 107, 110, 112
Variola (*see* Smallpox)
Vector, 177 (*see also* Cloning vector)
arthropod, 155
disease, 153–154, 159, 173
gene therapy, 130–132
vaccine (*see* Vaccine, fusion vector, gene vector, live vector)
Vertical transmission, 154
Vesicle, 60
Vesicular stomatitis virus, 3
Vidarabine, 105
Videx (*see* Didanosine)
Viracept (*see* Nelfinavir)
Viral load, 142
Viramune (*see* Nevirapine)
Virino, 48–49, 177
Virion, 1, 176
Viroid, 10, 44–46, 117
Virulence, 85
determinant, 86
moderation of, 155, 160
reversion to, 83–84, 86, 94
Virus X theory, 155–156
Virusoid, 45–46
Vistide (*see* Cidofovir)

War, 155, 164, 168
Western blotting, 63, 71, 136, 177
WIN compounds, 101

Xenotransplantation, 152, 165–166

Yeast, 86, 91, 124–125, 148
Yellow fever, 153
vaccine, 83

Zalcitabine, 106
Zerit (*see* Stavudine)
Zeta chain, 58
Zidovudine (*see* Azidothymidine)
Zoonosis, 152–156, 158–159, 165–167, 176
Zoster, 97, 100
Zovirax (*see* Aciclovir)